Stochastic Networks

Communication networks underpin our modern world, and provide fascinating and challenging examples of large-scale stochastic systems. Randomness arises in communication systems at many levels: for example, the initiation and termination times of calls in a telephone network, or the statistical structure of the arrival streams of packets at routers in the Internet. How can routing, flow control and connection acceptance algorithms be designed to work well in uncertain and random environments?

This compact introduction illustrates how stochastic models can be used to shed light on important issues in the design and control of communication networks. It will appeal to readers with a mathematical background wishing to understand this important area of application, and to those with an engineering background who want to grasp the underlying mathematical theory. Each chapter ends with exercises and suggestions for further reading.

FRANK KELLY is Professor of the Mathematics of Systems at the University of Cambridge. His main research interests are in random processes, networks and optimization. He is especially interested in applications to the design and control of networks and to the understanding of self-regulation in large-scale systems.

ELENA YUDOVINA is a postdoctoral research fellow at the University of Michigan. Her research interests are in applications of queueing theory. She received her PhD from the University of Cambridge, where her interest in the subject was sparked by a course on stochastic networks taught by Frank Kelly.

INSTITUTE OF MATHEMATICAL STATISTICS TEXTBOOKS

IMS Textbooks give introductory accounts of topics of current concern suitable for advanced courses at master's level, for doctoral students and for individual study. They are typically shorter than a fully developed textbook, often arising from material created for a topical course. Lengths of 100–290 pages are envisaged. The books typically contain exercises.

Stochastic Networks

FRANK KELLY
University of Cambridge

ELENA YUDOVINA
University of Michigan, Ann Arbor

CAMBRIDGE
UNIVERSITY PRESS

University Printing House, Cambridge CB2 8BS, United Kingdom

One Liberty Plaza, 20th Floor, New York, NY 10006, USA

477 Williamstown Road, Port Melbourne, VIC 3207, Australia

314-321, 3rd Floor, Plot 3, Splendor Forum, Jasola District Centre, New Delhi - 110025, India

79 Anson Road, #06-04/06, Singapore 079906

Cambridge University Press is part of the University of Cambridge.

It furthers the University's mission by disseminating knowledge in the pursuit of education, learning and research at the highest international levels of excellence.

www.cambridge.org
Information on this title: www.cambridge.org/9781107035775

First published 2014

A catalogue record for this publication is available from the British Library

ISBN 978-1-107-03577-5 Hardback
ISBN 978-1-107-69170-4 Paperback

Contents

Preface

Communication networks underpin our modern world, and provide fascinating and challenging examples of large-scale stochastic systems. Randomness arises in communication systems at many levels: for example, the initiation and termination times of calls in a telephone network, or the statistical structure of the arrival streams of packets at routers in the Internet. How can routing, flow control, and connection acceptance algorithms be designed to work well in uncertain and random environments? And can we design these algorithms using simple local rules so that they produce coherent and purposeful behaviour at the macroscopic level?

The first two parts of the book will describe a variety of classical models that can be used to help understand the behaviour of large-scale stochastic networks. Queueing and loss networks will be studied, as well as random access schemes and the concept of an effective bandwidth. Parallels will be drawn with models from physics, and with models of traffic in road networks.

The third part of the book will study more recently developed models of packet traffic and of congestion control algorithms in the Internet. This is an area of some practical importance, with network operators, content providers, hardware and software vendors, and regulators actively seeking ways of delivering new services reliably and effectively. The complex interplay between end-systems and the network has attracted the attention of economists as well as mathematicians and engineers.

We describe enough of the technological background to communication networks to motivate our models, but no more. Some of the ideas described in the book are finding application in financial, energy, and economic networks as computing and communication technologies transform these areas. But communication networks currently provide the richest and best-developed area of application within which to present a connected account of the ideas.

The lecture notes that have become this book were used for a Masters course ("Part III") in the Faculty of Mathematics at the University of Cambridge. This is a one-year postgraduate course which assumes a mathematically mature audience, albeit from a variety of different mathematical backgrounds. Familiarity with undergraduate courses on Optimization and Markov Chains is helpful, but not absolutely necessary. Appendices are provided on continuous time Markov processes, Little's law, Lagrange multipliers, and Foster–Lyapunov criteria, reviewing the material needed. At Cambridge, students may attend other courses where topics touched on in these notes, for example Poisson point processes, large deviations, or game theory, are treated in more depth, and this course can serve as additional motivation.

Suggestions on further reading are given at the end of each chapter; these include reviews where the historical development and attribution of results can be found – we have not attempted this here – as well as some recent papers which give current research directions.

The authors are grateful to students of the course at Cambridge (and, for one year, Stanford) for their questions and insights, and especially to Joe Hurd, Damon Wischik, Gaurav Raina, and Neil Walton, who produced earlier versions of these notes.

Frank Kelly and Elena Yudovina

Overview

This book is about stochastic networks and their applications. Large-scale systems of interacting components have long been of interest to physicists. For example, the behaviour of the air in a room can be described at the microscopic level in terms of the position and velocity of each molecule. At this level of detail a molecule's velocity appears as a random process. Consistent with this detailed microscopic description of the system is macroscopic behaviour best described by quantities such as temperature and pressure. Thus the pressure on the wall of the room is an average over an area and over time of many small momentum transfers as molecules bounce off the wall, and the relationship between temperature and pressure for a gas in a confined volume can be deduced from the microscopic behaviour of molecules.

Economists, as well as physicists, are interested in large-scale systems, driven by the interactions of agents with preferences rather than inanimate particles. For example, from a market with many heterogeneous buyers and sellers there may emerge the notion of a price at which the market clears.

Over the last 100 years, some of the most striking examples of large-scale systems have been technological in nature and constructed by us, from the telephony network through to the Internet. Can we relate the microscopic description of these systems in terms of calls or packets to macroscopic consequences such as blocking probabilities or throughput rates? And can we design the microscopic rules governing calls or packets to produce more desirable macroscopic behaviour? These are some of the questions we address in this book. We shall see that there are high-level constructs that parallel fundamental concepts from physics or economics such as energy or price, and which allow us to reason about the systems we design.

In this chapter we briefly introduce some of the models that we shall encounter. We'll see in later chapters that for the systems we ourselves construct, we are sometimes able to use simple local rules to produce macroscopic behaviour which appears coherent and purposeful.

Queueing and loss networks

We begin Chapter 1 with a brief introduction to Markov chains and Markov processes, which will be the underlying probabilistic model for most of the systems we consider. In Chapter 2 we study queueing networks, in which customers (or jobs or packets) wait for service by one or several servers.

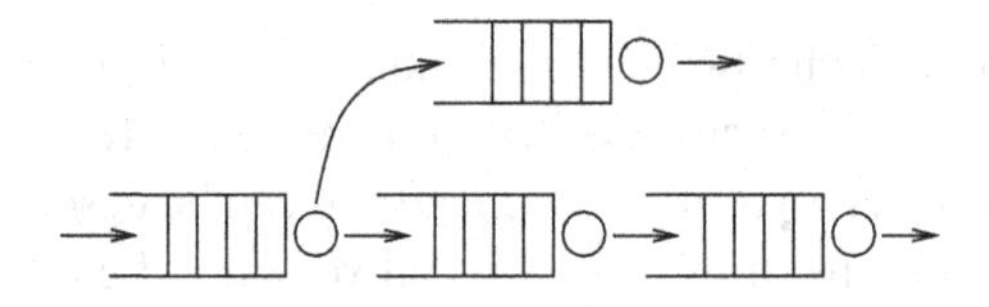

Figure 1 A network of four queues.

First, we look at a single queue. We shall see how to model it as a Markov process, and derive information on the distribution of the queue size. We then look briefly at a network of queues. Starting from a simplified description of each queue, we shall obtain information about the system behaviour. We define a traffic intensity for a simple queue, and identify Poisson flows in a network of queues.

A natural queueing discipline is first-come-first-served, and we also look at processor sharing, where the server shares its effort equally over all the customers present in the queue.

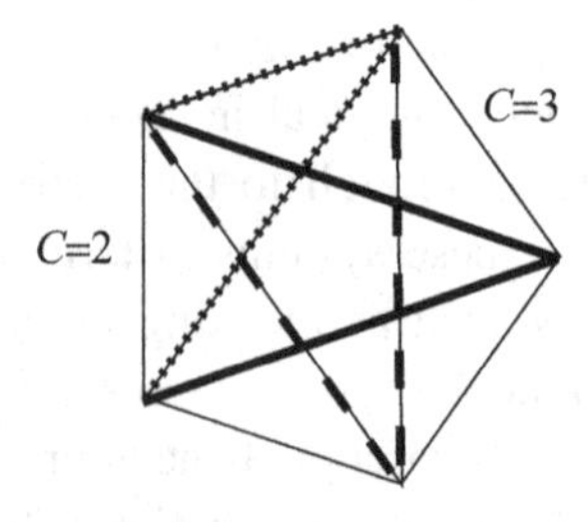

Figure 2 A loss network with some of the routes highlighted.

In Chapter 3, we move on to consider loss networks. A loss network consists of several links, each of which may have a number of circuits. The classical example of its use is to model landline telephone connections between several cities. When a telephone call is placed between one node and another, it needs to hold simultaneously a circuit on each of the links on a route between the two nodes – otherwise, the call is lost. Note the key differences between this and a queueing network, which are summarized in Table 1.

Table 1. *Queueing versus loss networks*

Queueing networks	Loss networks
Sequential use of resources	Simultaneous resource possession
Congestion $\Longrightarrow$ delay	Congestion $\Longrightarrow$ loss

First, we treat loss networks where routing is fixed in advance. We show that the simple rules for call acceptance lead to a stationary distribution for the system state that is centred on the solution to a certain optimization problem, and we'll relate this problem to a classical approximation procedure.

Next, we allow calls to be rerouted if they are blocked on their primary route. One example we consider is the following model of a loss network on a complete graph. Suppose that if a call arrives between two nodes and there is a circuit available on the direct link between the two nodes, the call is carried on the direct link. Otherwise, we attempt to redirect the call via another node (chosen at random), on a path through two links. (Figure 2 shows the single links and some of the two-link rerouting options.)

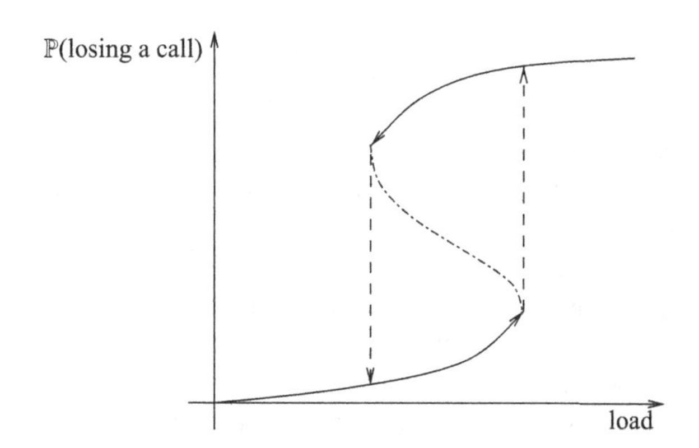

Figure 3 Hysteresis.

What is the loss probability in such a network as a function of the arrival rates? In Figure 3 we sketch the proportion of calls lost as the load on the system is varied. The interesting phenomenon that can occur is as follows: as the load is slowly increased, the loss probability increases slowly to begin with, but then jumps suddenly to a higher value; if the load is then slowly decreased, the loss probability does not trace the same curve, but drops back at a noticeably lower level of load. The system exhibits *hysteresis*, a phenomenon more often associated with magnetic materials.

How can this be? We model the network as an irreducible finite-state Markov chain, so it must have a *unique* stationary distribution, hence a *unique* probability of losing a call at a given arrival rate.

On the other hand, it makes sense. If the proportion of occupied circuits is low, a call is likely to be carried on the direct route through a single link; if the proportion of occupied circuits is high, more calls will be carried along indirect routes through two circuits, which may in turn keep link utilization high.

How can both of these insights be true? We'll see that the resolution concerns two distinct scaling regimes, obtained by considering either longer and longer time intervals, or larger and larger numbers of nodes.

We end Chapter 3 with a discussion of a simple dynamic routing strategy which allows a network to respond robustly to failures and overloads. We shall use this discussion to illustrate a more general phenomenon, *resource pooling*, that arises in systems where spare capacity in part of the network can be used to deal with excess load elsewhere. Resource pooling allows systems to operate more efficiently, but we'll see that this is sometimes at the cost of losing early warning signs that the system is about to saturate.

Decentralized optimization

A major practical and theoretical issue in the design of communication networks concerns the extent to which control can be decentralized, and in Chapter 4 we place this issue in a wider context through a discussion of some ideas from physics and economics.

In our study of loss networks we will have seen a network implicitly solving an optimization problem in a decentralized manner. This is reminiscent of various models in physics. We look at a very simple model of electron motion and establish Thomson's principle: the pattern of potentials in a network of resistors is just such that it minimizes heat dissipation for a given level of current flow. The local, random behaviour of the electrons results in the network as a whole solving a rather complex optimization problem.

Of course, simple local rules may lead to poor system behaviour if the rules are the wrong ones. We'll look at a simple model of a road traffic network to provide a chastening example of this. Braess's paradox describes how, if a new road is added to a congested network, the average speed of traffic may fall rather than rise, and indeed everyone's journey time may lengthen. It is possible to alter the local rules, by the imposition of appropriate tolls, so that the network behaves more sensibly, and indeed road traffic networks have long provided an example of the economic principle that externalities need to be appropriately penalized if the invisible hand is to lead to optimal behaviour.

In a queueing or loss network we can approach the optimization of the system at various levels, corresponding to the two previous models: we can dynamically route customers or calls according to the instantaneous state of the network locally, and we can also measure aggregate flows over longer periods of time, estimate externalities and use these estimates to decide where to add additional capacity.

Random access networks

In Chapter 5, we consider the following model. Suppose multiple base stations are connected with each other via a satellite, as in Figure 4. If a base station needs to pass a message to another station it sends it via the satellite, which then broadcasts the message to all of them (the address is part of the message, so it is recognized by the correct station at the end). If transmissions from two or more stations overlap, they interfere and need to be retransmitted. The fundamental issue here is contention resolution: how should stations arrange their transmissions so that at least some of them get through?

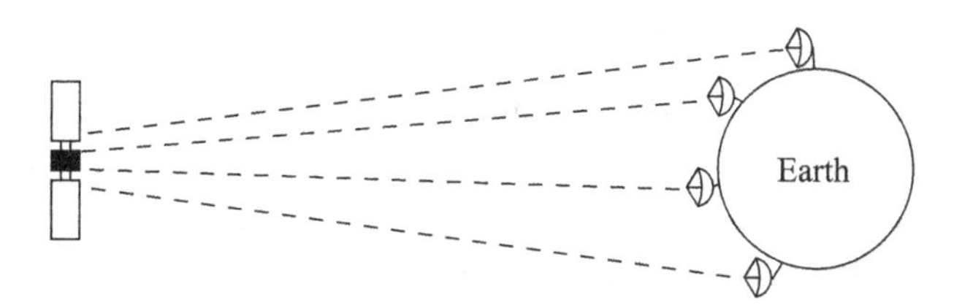

Figure 4 Multiple base stations in contact via a satellite.

If stations could instantaneously sense when another station is transmitting, there would be no collisions. The problem arises because the finite speed of light causes a delay between the time when a station starts transmitting and the time when other stations can sense this interference. As processing speeds increase, speed-of-light delays pose a problem over shorter and shorter distances, so that the issue now arises even for distances within a building or less.

Consider some approaches.

- We could divide time into slots, and assign, say, every fourth slot to each of the stations. However, this only works if we know how many stations there are and the load from each of them.
- We could implement a token ring: set up an order between the stations, and have the last thing that a station transmits be a "token" which means that it is done transmitting. Then the next station is allowed to start transmitting (or it may simply pass on the token). However, if there is a large number of stations, then simply passing the token around all of them will take a very long time.
- The ALOHA protocol: listen to the channel; if nobody else is transmitting, just start your own transmission. As it takes some time for messages to reach the satellite and be broadcast back, it is possible that there will be collisions (if two stations decide to start transmitting at sufficiently close times). If this happens, stop transmitting and wait for a *random* amount of time before trying to retransmit.
- The Ethernet protocol: after k unsuccessful attempts, wait for a random time with mean 2^k before retransmitting.

We shall study the last two approaches, which have advantages for the challenging case where the number of stations is large or unknown, each needing to access the channel infrequently.

We end Chapter 5 with a discussion of distributed random access, where each station can hear only a subset of the other stations.

Broadband networks

Consider a communication link of total capacity C, which is being shared by lots of connections, each of which needs a randomly fluctuating rate. We are interested in controlling the probability that the link is overloaded by controlling the admission of connections into the network.

If the rate is constant, as in the first panel of Figure 5, we could control admission as to a single-link loss network: don't admit another connection if the sum of the rates would exceed the capacity. But what should we do if the rate needed by a connection fluctuates, as in the second panel? A conservative approach might be to add up the peaks and not admit another

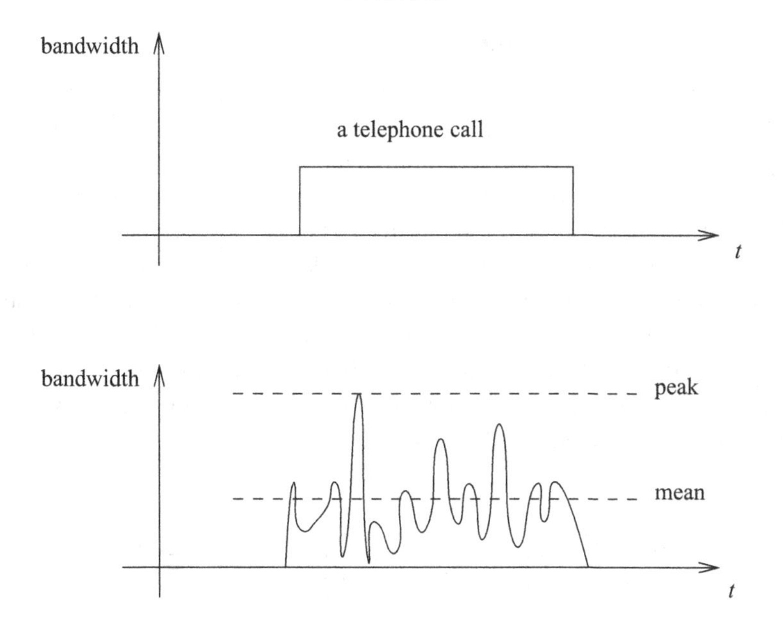

Figure 5 Possible bandwidth profiles; mean and peak rates.

connection if the sum would exceed the capacity. But this looks very conservative, and would waste a lot of capacity if the peak/mean ratio is high. Or we might add up the means and only refuse a connection if the sum of the means would exceed the capacity. If there is enough buffering available to smooth out the load, we might expect a queueing model to be stable, but connections would suffer long delays and this would not work for real-time communication.

In Chapter 6, we use a large deviations approach to define an *effective bandwidth* somewhere between the mean and the peak; this will depend on the statistical characteristics of the bandwidth profiles and on the desired low level of overload probability. Connection acceptance takes on a simple form: add the effective bandwidth of a new request to the effective bandwidths of connections already in progress and accept the new request if the sum satisfes a bound. This approach allows the insights available from our earlier model of a loss network to transfer to the case where the bandwidth required by a connection fluctuates.

If the connections are transferring a file, rather than carrying a real-time video conversation, another approach is possible which makes more efficient use of capacity, as we describe next.

Internet modelling

What happens when I want to see a web page? My device (e.g. computer, laptop, phone) sends a request to the server, which then starts sending the page to me as a stream of packets through the very large network that is the Internet. The server sends one packet at first; my device sends an acknowledgement packet back; as the server receives acknowledgements it sends further packets, at an increasing rate. If a packet is lost (which will happen if a buffer within the network overflows) it is detected by the endpoints (packets have sequence numbers on them) and the server slows down. The process is controlled by TCP, the *transmission control protocol* of the Internet, implemented at the endpoints (on my device and on the server) – the network itself is essentially dumb.

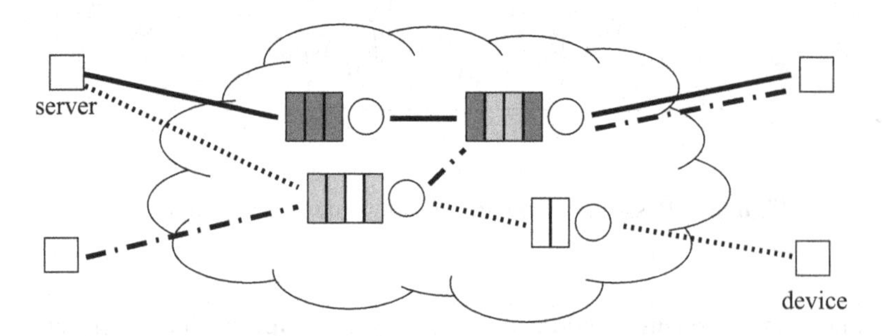

Figure 6 A schematic diagram of the Internet. Squares correspond to devices and servers, the network contains resources with buffers, and many flows traverse the network.

Even this greatly simplified model of the Internet raises many fascinating questions. Each flow is controlled by its own feedback loop, and these control loops interact with each other through their competition for shared resources. Will the network be stable, and what does this mean? How will the network end up sharing the resources between the flows?

In Chapter 7, we discuss various forms of fairness and their interpretation in terms of optimization problems. We will study dynamical models of congestion control algorithms that correspond to primal or dual approaches to the solution of these optimization problems.

Over longer time scales, flows will come and go, as web pages and files are requested or their transfers are completed. On this time scale, the network's overall behaviour resembles a processor-sharing queueing system, with congestion decreasing flow rates and lengthening the delay before a flow is completed. We'll look at models of this behaviour in Chapter 8.

We shall see that to understand the Internet's behaviour we need to consider many time and space scales, depending on the level of aggregation and the time intervals of interest. At the packet level the system looks like a queueing network, and we need to model packet-level congestion control algorithms and scheduling disciplines; but if we look on a longer time scale, a flow makes simultaneous use of resources along its route, and flow rates converge to the solution of an optimization problem; and over longer time scales still, where the numbers of flows can fluctuate significantly, the system behaves as a processor-sharing queue.

Part I

1

Markov chains

Our fundamental model will be a Markov chain. We do not presume to give an introduction to the theory of Markov chains in this chapter; for that, see Norris (1998). In this chapter, we briefly review the basic definitions.

1.1 Definitions and notation

Let $\mathcal{S}$ be a countable state space, and let $P = (p(i, j), i, j \in \mathcal{S})$ be a matrix of transition probabilities. That is, $p(i, j) \geq 0$ for all i, j, and $\sum_j p(i, j) = 1$. A collection of random variables $(X(n), n \in \mathbb{Z}_+)$, defined on a common probability space and taking values in $\mathcal{S}$, is a *Markov chain* with transition probabilities P if, for $j, k \in \mathcal{S}$ and $n \in \mathbb{Z}_+$, we have

$$\mathbb{P}(X(n+1) = k \mid X(n) = j, X(n-1) = x_{n-1}, \ldots, X(0) = x_0) = p(j, k)$$

whenever the conditioning event has positive probability. That is, the conditional probability depends only on j and k, and not on the earlier states $x_0, \ldots, x_{n-1}$, or on n (the *Markov property*). Here we have used capital letters to denote random variables, and lower case letters to refer to specific states in $\mathcal{S}$. Thus, $X(0)$ is a random variable, and x_0 is a particular element of $\mathcal{S}$.

We assume that any Markov chain with which we deal is irreducible, and often we additionally assume aperiodicity. A Markov chain is *irreducible* if every state in $\mathcal{S}$ can be reached, directly or indirectly, from every other state. A Markov chain is *periodic* if there exists an integer $\delta > 1$ such that $\mathbb{P}(X(n+\tau) = j \mid X(n) = j) = 0$ unless τ is divisible by δ; otherwise the chain is *aperiodic*.

A Markov chain evolves in discrete time, potentially changing state at integer times. Next we consider continuous time. Let $Q = (q(i, j), i, j \in \mathcal{S})$ be a matrix with $q(i, j) \geq 0$ for $i \neq j$, $q(i, i) = 0$, and $0 < q(i) \equiv \sum_{j \in S} q(i,j) < \infty$ for all i. Let

$$p(j,k) = \frac{q(j,k)}{q(j)}, \qquad j,k \in \mathcal{S},$$

so that $P = (p(i,j), i, j \in \mathcal{S})$ is a matrix of transition probabilities. Informally, a Markov process in continuous time rests in state i for a time that is exponentially distributed with parameter $q(i)$ (hence mean $q(i)^{-1}$), and then jumps to state j with probability $p(i,j)$. Note that $p(i,i) = 0$ in our definition, so the Markov process must change state when it jumps.

Formally, let $(X^J(n), n \in \mathbb{Z}_+)$ be a Markov chain with transition probabilities P, and let $T_0, T_1, \ldots$ be an independent sequence of independent exponentially distributed random variables with unit mean. Now let $S_n = q(X^J(n))^{-1} T_n$; thus if $X^J(n) = i$ then S_n is an exponentially distributed random variable with parameter $q(i)$. (The letter S stands for "sojourn time".) We shall define a process evolving in continuous time by using the Markov chain X^J to record the sequence of states occupied, and the sequence S_n to record the time spent in successive states. To do this, let

$$X(t) = X^J(N) \quad \text{for} \quad S_0 + \ldots + S_{N-1} \le t < S_0 + \ldots + S_N.$$

Then it can be shown that the properties of the Markov chain X^J together with the memoryless property of the exponential distribution imply that, for $t_0 < t_1 < \ldots < t_n < t_{n+1}$,

$$\begin{aligned}\mathbb{P}(X(t_{n+1}) = x_{n+1} \mid X(t_n) = x_n, X(t_{n-1}) = x_{n-1}, \ldots, X(t_0) = x_0)\\ = \mathbb{P}(X(t_{n+1}) = x_{n+1} \mid X(t_n) = x_n)\end{aligned}$$

whenever the event $\{X(t_n) = x_n\}$ has positive probability; and, further, that for any pair $j \neq k \in \mathcal{S}$

$$\lim_{\delta \to 0} \frac{\mathbb{P}(X(t+\delta) = k \mid X(t) = j)}{\delta} = q(j,k)$$

exists and depends only on j and k, whenever the event $\{X(t) = j\}$ has positive probability. Intuitively, we can think of conditioning on the entire trajectory of the process up to time t – we claim that the only important information is the state at time t, not how the process got there. We call $q(j,k)$ the *transition rate* from j to k. The Markov chain X^J is called the *jump chain*. The sum $\sum_0^\infty S_n$ may be finite, in which case an infinite number of jumps takes place in a finite time (and the process "runs out of instructions"); if not, we have defined the *Markov process* $(X(t), t \in \mathbb{R}_+)$.

We shall only be concerned with countable state space processes. There is a rich theory of Markov chains/processes defined on an uncountable state space; in some other texts the term "Markov process" refers to continuous space, rather than continuous time. Even with only a countable state space it is possible to construct Markov processes with strange behaviour. These will be excluded: we shall assume that all our Markov processes are irreducible, remain in each state for a positive length of time, and are incapable of passing through an infinite number of states in finite time. See Appendix A for more on this.

A Markov chain or process may possess an *equilibrium distribution*, i.e. a collection $\pi = (\pi(j), j \in \mathcal{S})$ of positive numbers summing to unity that satisfy

$$\pi(j) = \sum_{k \in \mathcal{S}} \pi(k) p(k, j), \qquad \forall j \in \mathcal{S}, \tag{1.1}$$

for a Markov chain, or

$$\pi(j) \sum_{k \in \mathcal{S}} q(j, k) = \sum_{k \in \mathcal{S}} \pi(k) q(k, j), \qquad \forall j \in \mathcal{S}, \tag{1.2}$$

for a Markov process (the *equilibrium equations*).

Under the assumptions we have made for a Markov process, if π exists then it will be unique. It will then also be the limiting, ergodic, and stationary distribution:

Limiting $\forall j \in \mathcal{S}, \mathbb{P}(X(t) = j) \to \pi(j)$ as $t \to \infty$.

Ergodic $\forall j \in \mathcal{S}, \frac{1}{T} \int_0^T I\{X(t) = j\}\, dt \to \pi(j)$ as $T \to \infty$ with probability 1. (The integral is the amount of time, between 0 and T, that the process spends in state j.)

Stationary If $\mathbb{P}(X(0) = j) = \pi(j)$ for all $j \in \mathcal{S}$, then $\mathbb{P}(X(t) = j) = \pi(j)$ for all $j \in \mathcal{S}$ and all $t \geq 0$.

If an equilibrium distribution does not exist then $\mathbb{P}(X(t) = j) \to 0$ as $t \to \infty$ for all $j \in \mathcal{S}$. An equilibrium distribution will not exist if we can find a collection of positive numbers satisfying the equilibrium equations whose sum is infinite. When the state space $\mathcal{S}$ is finite, an equilibrium distribution will always exist. All of this paragraph remains true for Markov chains, except that for a periodic chain there may not be convergence to a limiting distribution.

A chain or process for which $\mathbb{P}(X(0) = j) = \pi(j)$ for all $j \in \mathcal{S}$ is termed *stationary*. A stationary chain or process can be defined for all $t \in \mathbb{Z}$ or $\mathbb{R}$ – we imagine time has been running from $-\infty$. We shall often refer to a stationary Markov process as being in equilibrium.

Exercises

Exercise 1.1 Let S be an exponentially distributed random variable. Show that for any $s, t > 0$

$$\mathbb{P}(S > s + t \mid S > t) = \mathbb{P}(S > s),$$

the *memoryless property* of the exponential distribution.

Let S, T be independent random variables having exponential distributions with parameters λ, μ. Show that

$$Z = \min\{S, T\}$$

also has an exponential distribution, with parameter $\lambda + \mu$.

Exercise 1.2 A Markov process has transition rates $(q(j,k),\ j,k \in \mathcal{S})$ and equilibrium distribution $(\pi(j),\ j \in \mathcal{S})$. Show that its jump chain has equilibrium distribution $(\pi^J(j),\ j \in \mathcal{S})$ given by

$$\pi^J(j) = G^{-1}\pi(j)q(j)$$

provided

$$G = \sum_j \sum_k \pi(j)q(j,k) < \infty. \tag{1.3}$$

Exercise 1.3 Show that the Markov property is equivalent to the following statement: for any $t_{-k} < \ldots < t_{-1} < t_0 < t_1 < \ldots < t_n$,

$$\begin{aligned}\mathbb{P}(X(t_n) = x_n, X(t_{n-1}) = x_{n-1}, \ldots, X(t_{-k}) = x_{-k} \mid X(t_0) = x_0) \\ = \mathbb{P}(X(t_n) = x_n, \ldots, X(t_1) = x_1 \mid X(t_0) = x_0) \\ \times \mathbb{P}(X(t_{-1}) = x_{-1}, \ldots, X(t_{-k}) = x_{-k} \mid X(t_0) = x_0),\end{aligned}$$

whenever the conditioning event has positive probability. That is, "conditional on the present, the past and the future are independent".

1.2 Time reversal

For a stationary chain or process, we can consider what happens if we run time backwards. Because the Markov property is symmetric in time (Exercise 1.3), the *reversed* chain or process will again be Markov. Let's calculate its transition rates.

Proposition 1.1 *Suppose $(X(t), t \in \mathbb{R})$ is a stationary Markov process with transition rates $(q(i,j),\ i,j \in \mathcal{S})$ and equilibrium distribution $\pi = (\pi(j), j \in \mathcal{S})$. Let $Y(t) = X(-t)$, $t \in \mathbb{R}$. Then $(Y(t), t \in \mathbb{R})$ is a stationary*

Markov process with the same equilibrium distribution π and transition rates

$$q'(j,k) = \frac{\pi(k)q(k,j)}{\pi(j)}, \qquad \forall j,k \in \mathcal{S}.$$

Proof We know $(Y(t), t \in \mathbb{R})$ is a Markov process; we determine its transition rates by an application of Bayes' theorem:

$$\begin{aligned}
\mathbb{P}(Y(t+\delta) = k \mid Y(t) = j) &= \frac{\mathbb{P}(Y(t+\delta) = k, Y(t) = j)}{\mathbb{P}(Y(t) = j)} \\
&= \frac{\mathbb{P}(X(-t-\delta) = k, X(-t) = j)}{\pi(j)} \\
&= \frac{\pi(k)\mathbb{P}(X(-t) = j \mid X(-(t+\delta)) = k)}{\pi(j)} \\
&= \frac{\pi(k)(q(k,j)\delta + o(\delta))}{\pi(j)}.
\end{aligned}$$

Now divide both sides by δ and let $\delta \to 0$ to obtain the desired expression for $q'(j,k)$.

Of course, $\mathbb{P}(Y(t) = i) = \pi(i)$ for all $i \in \mathcal{S}$ and for all t, and so π is the stationary distribution. □

If the reversed process has the same transition rates as the original process we call the process *reversible*. In order for this to hold, i.e. to have $q(j,k) = q'(j,k)$, we need the following *detailed balance* equations to be satisfied:

$$\pi(j)q(j,k) = \pi(k)q(k,j), \quad \forall j,k \in \mathcal{S}. \tag{1.4}$$

Detailed balance says that, in equilibrium, transitions from j to k happen "as frequently" as transitions from k to j.

Note that detailed balance implies the equilibrium equations (1.2), which are sometimes known as *full balance*. When it holds, detailed balance is much easier to check than full balance. Consequently, we will often look for equilibrium distributions of Markov processes by trying to solve detailed balance equations.

Exercises

Exercise 1.4 Check that the distribution $(\pi(j), j \in \mathcal{S})$ and the transition rates $(q'(j,k), j,k \in \mathcal{S})$ found in Proposition 1.1 satisfy the equilibrium equations.

Establish the counterpart to Proposition 1.1 for a Markov chain.

Exercise 1.5 Associate a graph with a Markov process by letting an edge join states j and k if either $q(j,k)$ or $q(k,j)$ is positive. Show that if the graph associated with a stationary Markov process is a tree, the process is reversible.

1.3 Erlang's formula

In the early twentieth century Agner Krarup Erlang worked for the Copenhagen Telephone Company, and he was interested in calculating how many parallel circuits were needed on a telephone link to provide service (say, out of a single town, or between two cities). We will later look at generalizations of this problem for networks of links; right now, we will look at a single link.

Suppose that calls arrive to a link as a Poisson process of rate λ. The link has C parallel circuits, and while a call lasts it uses one of the circuits. We will assume that each call lasts for an exponentially distributed amount of time with parameter μ, and that these call holding periods are independent of each other and of the arrival times. If a call arrives to the link and finds that all C circuits are busy, it is simply lost. We would like to estimate the probability that an arriving call is lost.

Let $X(t) = j$ be the number of busy circuits on the link. This is a Markov process, with transition rates given by

$$q(j, j+1) = \lambda, \; j = 0, 1, \ldots, C-1; \qquad q(j, j-1) = j\mu, \; j = 1, 2, \ldots, C.$$

The upward transition is equivalent to the assumption that the arrival process is Poisson of rate λ; the downward transition rate arises since we are looking for the first of j calls to finish – recall Exercise 1.1.

We try to solve the detailed balance equations:

$$\pi(j-1)q(j-1, j) = \pi(j)q(j, j-1)$$
$$\Longrightarrow \pi(j) = \frac{\lambda}{\mu j}\pi(j-1) = \cdots = \left(\frac{\lambda}{\mu}\right)^j \frac{1}{j!}\pi(0).$$

Adding the additional constraint $\sum_{j=0}^{C} \pi(j) = 1$, we find

$$\pi(0) = \left(\sum_{j=0}^{C} \left(\frac{\lambda}{\mu}\right)^j \frac{1}{j!}\right)^{-1}.$$

This is the equilibrium probability of all circuits being free.

The equilibrium probability that all circuits are busy is therefore

$$\pi(C) = E\left(\frac{\lambda}{\mu}, C\right), \qquad \text{where } E(\nu, C) = \frac{\nu^C/C!}{\sum_{j=0}^{C} \nu^j/j!}. \tag{1.5}$$

This is known as *Erlang's formula.*

This was a model that assumed that the arrival rate of calls is constant at all times. Suppose, however, that we have a finite population of M callers, so that those already connected will not be trying to connect again, and suppose someone not already connected calls at rate η. We consider a link with C parallel circuits, where $C < M$.

Letting $X(t) = j$ be the number of busy lines, we now have the transition rates

$$q(j, j+1) = \eta(M-j),\ j = 0, \dots, C-1; \qquad q(j, j-1) = \mu j,\ j = 1, 2, \dots, C.$$

The equilibrium distribution $\pi_M(j)$ now satisfies (check this!)

$$\pi_M(j-1)q(j-1, j) = \pi_M(j)q(j, j-1)$$
$$\implies \pi_M(j) = \pi_M(0)\binom{M}{j}\left(\frac{\eta}{\mu}\right)^j.$$

Next we find the distribution of the number of busy circuits when a new call is initiated. Now

$$\mathbb{P}(j \text{ lines are busy and a call is initiated in } (t, t+\delta))$$
$$= \pi_M(j)(\eta(M-j)\delta + o(\delta)), \qquad j = 0, 1, \dots, C,$$

and the probability of a call being initiated is just the sum of these terms over all j. Letting $\delta \to 0$, we find that the conditional probability of j lines being busy when a new call arrives is

$$\frac{\pi_M(j)\eta(M-j)}{\sum_i \pi_M(i)\eta(M-i)} \propto \pi_M(j)(M-j)$$
$$\propto \binom{M}{j}(M-j)\left(\frac{\eta}{\mu}\right)^j$$
$$\propto \binom{M-1}{j}\left(\frac{\eta}{\mu}\right)^j.$$

The symbol "$\propto$" means "is proportional to". We have used the trick of only leaving the terms that depend on j; the normalization constant can be found later, from the fact that we have a probability distribution that adds up to 1. If we now enforce this condition, we see

$$\mathbb{P}(j \text{ lines are busy when a call is initiated}) = \pi_{M-1}(j), \qquad j = 0, 1, \dots, C.$$

Thus

$$\mathbb{P}(\text{an arriving call is lost}) = \pi_{M-1}(C) < \pi_M(C);$$

an arriving call sees a distribution of busy lines that is not the time-averaged distribution of busy lines, π_M.

Note that if $M \to \infty$ with ηM held fixed at λ, we recover the previous model, and the probability that an arriving call finds all lines busy approaches the time-averaged probability, given by Erlang's formula. This is an example of a more general phenomenon, called the *PASTA* property ("Poisson arrivals see time averages").

Exercises

Exercise 1.6 A Markov process has transition rates $(q(j,k),\ j,k \in \mathcal{S})$ and equilibrium distribution $(\pi(j),\ j \in \mathcal{S})$. A Markov chain is formed by observing the transitions of this process: at the successive jump times of the process, the state j just before, and the state k just after, the jump are recorded as an ordered pair $(j,k) \in \mathcal{S}^2$. Write down the transition probabilities of the resulting Markov chain, and show that it has equilibrium distribution

$$\pi'(j,k) = G^{-1}\pi(j)q(j,k)$$

provided (1.3) holds.

Give an alternative interpretation of $\pi'(j,k)$ in terms of the conditional probability of seeing the original process jump from state j to state k in the interval $(t, t+\delta)$, given that the process is in equilibrium and a jump occurs in that interval.

Exercise 1.7 Show that the mean number of circuits in use in the model leading to Erlang's formula, i.e. the mean of the equilibrium distribution π, is

$$\nu(1 - E(\nu, C)).$$

Exercise 1.8 Show that

$$\frac{d}{d\nu}E(\nu, C) = -(1 - E(\nu, C))(E(\nu, C) - E(\nu, C - 1)).$$

Exercise 1.9 Car parking spaces are labelled $n = 1, 2, \ldots$, where the label indicates the distance (in car lengths) the space is from a shop, and an arriving car parks in the lowest numbered free space. Cars arrive as a Poisson process of rate ν; parking times are exponentially distributed with unit mean, and are independent of each other and of the arrival process. Show that the distance at which a newly arrived car parks has expectation

$$\sum_{C=0}^{\infty} E(\nu, C),$$

where $E(\nu, C)$ is Erlang's formula.
[*Hint:* Calculate the probability the first C spaces are occupied.]

1.4 Further reading

Norris (1998) is a lucid and rigorous introduction to discrete and continuous time Markov chains. The fourth chapter of Whittle (1986) and the first chapter of Kelly (2011) give more extended discussions of reversibility than we have time for here.

2

Queueing networks

In this chapter we look at simple models of a queue and of a network of queues. We begin by studying the departure process from a queue with Poisson arrivals and exponential service times. This will be important to understand if customers leaving the queue move on to another queue in a network.

2.1 An $M/M/1$ queue

Suppose that the stream of customers arriving at a queue (the arrival process) forms a Poisson process of rate λ. Suppose further there is a single server and that customers' service times are independent of each other and of the arrival process and are exponentially distributed with parameter μ. Such a system is called an $M/M/1$ queue, the M's indicating the memoryless (exponential) character of the interarrival and service times, and the final digit indicating the number of servers. Let $X(t) = j$ be the number of customers in the queue, including the customer being served. Then it follows from our description of the queue that X is a Markov process with transition rates

$$q(j, j+1) = \lambda, \qquad j = 0, 1, \ldots,$$
$$q(j, j-1) = \mu, \qquad j = 1, 2, \ldots .$$

If the arrival rate λ is less than the service rate μ, the distribution

$$\pi(j) = (1-\rho)\rho^j, \qquad j = 0, 1, \ldots,$$

satisfies the detailed balance equations (1.4) and is thus the equilibrium distribution, where $\rho = \lambda/\mu$ is the *traffic intensity*.

Let $(X(t), t \in \mathbb{R})$ be a stationary $M/M/1$ queue. Since it is reversible, it is indistinguishable from the same process run backwards in time:

$$(X(t), t \in \mathbb{R}) \stackrel{\mathcal{D}}{=} (X(-t), t \in \mathbb{R}),$$

where $\stackrel{\mathcal{D}}{=}$ means *equal in distribution*. That is, there is no statistic that can distinguish the two processes.

Now consider the sample path of the queue. We can represent it in terms of two point processes (viewed here as sets of random times): the arrival process A, and the departure process D.

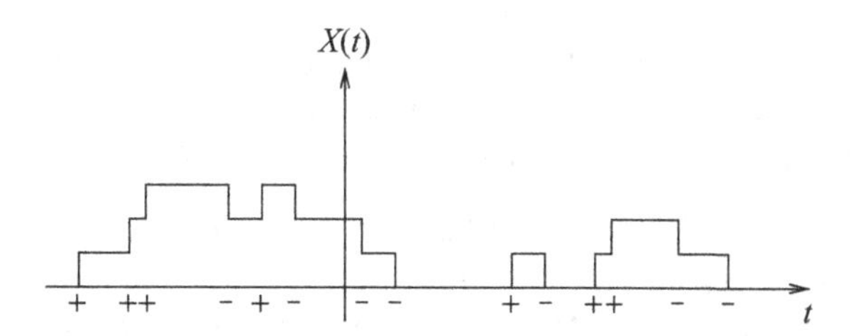

Figure 2.1 Sample path of the $M/M/1$ queue; + marks the points of A, − marks the points of D.

The arrival process A records the jumps up of the Markov process $X(t)$; the departure process D records the jumps down of the Markov process $X(t)$. By assumption, A is a Poisson process of rate λ. On the other hand, A is defined on $(X(t), t \in \mathbb{R})$ exactly as D is defined on $(X(-t), t \in \mathbb{R})$ – the departures of the original process are the arrivals of the reversed process. Since $(X(t), t \in \mathbb{R})$ and $(X(-t), t \in \mathbb{R})$ are distributionally equivalent, we conclude that D is also a Poisson process of rate λ. That is, we have shown the following theorem.

Theorem 2.1 (Burke, Reich) *In equilibrium, the departure process from an $M/M/1$ queue is a Poisson process.*

Remark 2.2 Time reversibility for the systems we consider holds only in equilibrium! For example, if we condition on the queue size being positive, then the time until the next departure is again exponential, but now with parameter μ rather than λ. A related comment is that A and D are both Poisson processes, but they are by no means independent.

To investigate the dependence structure, we introduce the following notation. For random variables or processes B_1 and B_2, we will write $B_1 \perp\!\!\!\perp B_2$ to mean that B_1 and B_2 are independent. Now, for a fixed time $t_0 \in \mathbb{R}$, the state of the queue up to time t_0 is independent of the future arrivals:

$$(X(t), t \le t_0) \perp\!\!\!\perp (A \cap (t_0, \infty)).$$

Applying this to the reversed process, we get for every fixed time $t_1 \in \mathbb{R}$

$$(X(t), t \ge t_1) \perp\!\!\!\perp (D \cap (-\infty, t_1)).$$

In particular, when the queue is in equilibrium, the number of people that are present in the queue at time t_1 is independent of the departure process up to time t_1. (But clearly it isn't independent of the future departure process!)

Exercises

Exercise 2.1 Upon an $M/M/1$ queue is imposed the additional constraint that arriving customers who find N customers already present leave and never return. Find the equilibrium distribution of the queue.

Exercise 2.2 An $M/M/2$ queue has two identical servers at the front. Find a Markov chain representation for it, and determine the equilibrium distribution of a stationary $M/M/2$ queue. Is the queue reversible?
[*Hint:* You can still use the number of customers in the queue as the state of the Markov chain.]

Deduce from the equilibrium distribution that the proportion of time both servers are idle is

$$\pi(0) = \frac{1-\rho}{1+\rho}, \qquad \text{where } \rho = \frac{\lambda}{2\mu} < 1.$$

Exercise 2.3 An $M/M/1$ queue has arrival rate ν and service rate μ, where $\rho = \nu/\mu < 1$. Show that the sojourn time (= queueing time + service time) of a typical customer is exponentially distributed with parameter $\mu - \nu$.

Exercise 2.4 Consider an $M/M/\infty$ queue with servers numbered $1, 2, \ldots$ On arrival a customer chooses the lowest numbered server that is free. Calculate the equilibrium probability that server j is busy.
[*Hint:* Compare with Exercise 1.9, and calculate the expected number out of the first C servers that are busy.]

2.2 A series of $M/M/1$ queues

The most obvious application of Theorem 2.1 is to a series of J queues arranged so that when a customer leaves a queue she joins the next one, until she has passed through all the queues, as illustrated in Figure 2.2.

Suppose the arrival stream at queue 1 is Poisson at rate ν, and that service times at queue j are exponentially distributed with parameter μ_j, where $\nu < \mu_j$ for $j = 1, 2, \ldots, J$. Suppose further that service times are independent of each other, including those of the same customer in different queues, and of the arrival process at queue 1.

By the analysis in Section 2.1, if we look at the system in equilibrium,

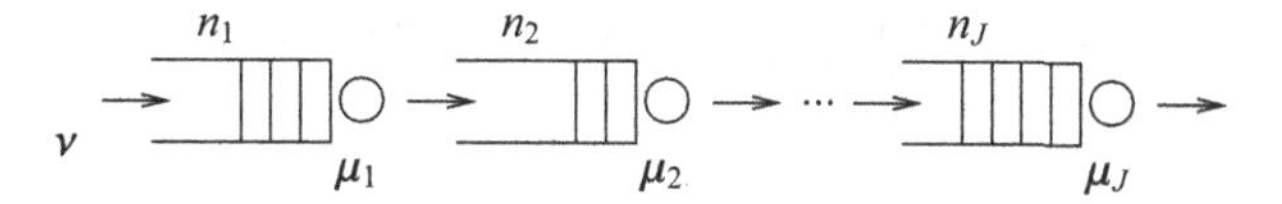

Figure 2.2 A series of $M/M/1$ queues.

then (by induction) each of the queues has a Poisson arrival process, so this really is a series of $M/M/1$ queues. Let $n(t) = (n_1(t), \ldots, n_J(t))$ be the Markov process giving the number of customers in each of the queues at time t. In equilibrium we know the marginal distribution of the components:

$$\pi_j(n_j) = (1 - \rho_j)\rho_j^{n_j}, \qquad n_j = 0, 1, \ldots$$

where $\rho_j = \nu/\mu_j$. But what is the joint distribution of all J queue lengths? Equivalently, what is the dependence structure between them? To discover this, we develop a little further the induction that gave us the marginal distributions.

For a fixed time t_0, consider the following quantities:

(1) $n_1(t_0)$;
(2) departures from queue 1 prior to t_0;
(3) service times of customers in queues $2, 3, \ldots, J$;
(4) the remaining vector $(n_2(t_0), \ldots, n_J(t_0))$.

In Section 2.1, we have established $(1) \perp\!\!\!\perp (2)$. Also, $(3) \perp\!\!\!\perp (1, 2)$ by construction: the customers' service times at queues $2, \ldots, J$ are independent of the arrival process at queue 1 and service times there. Therefore, (1), (2) and (3) are mutually independent, and in particular $(1) \perp\!\!\!\perp (2, 3)$. On the other hand, (4) is a function of (2,3), so we conclude that $(1) \perp\!\!\!\perp (4)$.

Similarly, for each j we have $n_j(t_0) \perp\!\!\!\perp (n_{j+1}(t_0), \ldots, n_J(t_0))$. Therefore the equilibrium distribution $\pi(n)$ factorizes:

$$\pi(n_1, \ldots, n_J) = \prod_{j=1}^{J} \pi_j(n_j).$$

You can use this technique to show a number of other results for a series of $M/M/1$ queues; for example, Exercise 2.5 shows that the sojourn times of a customer in successive queues are independent. However, the technique is fragile; in particular, it does not allow a customer to leave a later queue and return to an earlier one. We will next develop a set of tools that does not give such fine-detail results, but can tolerate more general flow patterns.

Exercise

Exercise 2.5 Suppose that customers at each queue in a stationary series of $M/M/1$ queues are served in the order of their arrival. Note that, from a sample path such as that illustrated in Figure 2.1, each arrival time can be matched to a departure time, corresponding to the same customer. Argue that the sojourn time of a customer in queue 1 is independent of departures from queue 1 prior to her departure. Deduce that in equilibrium the sojourn times of a customer at each of the J queues are independent.

2.3 Closed migration processes

In this section, we will analyze a generalization of the series of queues example. It is simplest to just give a Markovian description. The state space

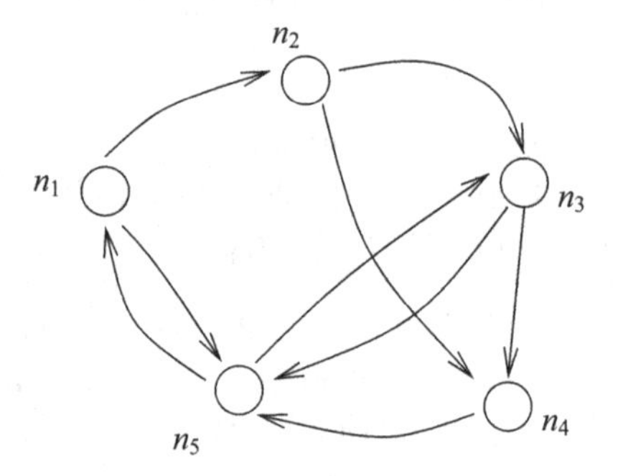

Figure 2.3 Closed migration process.

of the Markov process is $S = \{n \in \mathbb{Z}_+^J : \sum_{j=1}^J n_j = N\}$. Each state is written as $n = (n_1, \ldots, n_J)$, where n_j is the number of *individuals* in *colony* j.

For two different colonies j and k, define the operator T_{jk} as

$$T_{jk}(n_1, \ldots, n_J) = \begin{cases} (n_1, \ldots, n_j - 1, \ldots, n_k + 1, \ldots, n_J), & j < k, \\ (n_1, \ldots, n_k + 1, \ldots, n_j - 1, \ldots, n_J), & j > k. \end{cases}$$

That is, T_{jk} transfers one individual from colony j to colony k.

We now describe the rate at which transitions occur in our state space. We will only allow individuals to move one at a time, so transitions can only occur between a state n and $T_{jk}n$ for some j, k. We will assume that the transition rates have the form

$$q(n, T_{jk}(n)) = \lambda_{jk}\phi_j(n_j), \quad \phi_j(0) = 0.$$

That is, it is possible to factor the rate into a product of two functions: one depending only on the two colonies j and k, and another depending only on the number of individuals in the "source" colony j.

We will suppose that n is irreducible in S (in particular, that it is possible for individuals to get from any colony to any other colony, possibly in several steps). In this case, we call n a *closed migration process*.

We can model an s-server queue at colony j by taking $\phi_j(n) = \min(n, s)$. Each of the customers requires an exponential service time with parameter $\lambda_j = \sum_k \lambda_{jk}$; and once service is completed, the individual goes to colony k with probability λ_{jk}/λ_j.

Another important example is $\phi_j(n) = n$ for all j. These are the transition rates we get if individuals move independently of one another. This can be thought of as a network of *infinite-server* queues (corresponding to $s = \infty$ above), and is an example of a *linear migration process*; we study these further in Section 2.6. If $N = 1$, the single individual performs a random walk on the set of colonies, with equilibrium distribution (α_j), where the α_j satisfy

$$\alpha_j > 0, \qquad \sum_j \alpha_j = 1,$$

$$\alpha_j \sum_k \lambda_{jk} = \sum_k \alpha_k \lambda_{kj}, \qquad j = 1, 2, \dots, J. \tag{2.1}$$

We refer to these equations as the *traffic equations*, and we use them to define the quantities (α_j) in terms of (λ_{jk}) for a general closed migration process.

Remark 2.3 Note that the quantities λ_{jk} and $\phi_j(\cdot)$ are only well defined up to a constant factor. In particular, the (α_j) are only well defined after we have picked the particular set of (λ_{jk}).

Theorem 2.4 *The equilibrium distribution for a closed migration process is*

$$\pi(n) = G_N^{-1} \prod_{j=1}^{J} \frac{\alpha_j^{n_j}}{\prod_{r=1}^{n_j} \phi_j(r)}, \qquad n \in S.$$

Here, G_N is a normalizing constant, chosen so the distribution sums to 1, and (α_j) are the solution to the traffic equations (2.1).

Remark 2.5 Although this expression looks somewhat complicated, its form is really quite simple: the joint distribution factors as a product over individual colonies.

Proof In order to check that this is the equilibrium distribution, it suffices to verify that for each n the full balance equations hold:

$$\pi(n) \sum_j \sum_k q(n, T_{jk}n) \stackrel{?}{=} \sum_j \sum_k \pi(T_{jk}n) q(T_{jk}n, n).$$

These will be satisfied provided the following set of *partial balance* equations hold:

$$\pi(n) \sum_k q(n, T_{jk}n) \stackrel{?}{=} \sum_k \pi(T_{jk}n) q(T_{jk}n, n), \qquad \forall j.$$

That is, it suffices to check that, from any state, the rate of individuals leaving a given colony j is the same as the rate of individuals arriving into it. We now recall

$$q(n, T_{jk}n) = \lambda_{jk} \phi_j(n_j), \qquad q(T_{jk}n, n) = \lambda_{kj} \phi_k(n_k + 1),$$

and from the claimed form for π we have that

$$\pi(T_{jk}n) = \pi(n) \frac{\phi_j(n_j)}{\alpha_j} \frac{\alpha_k}{\phi_k(n_k + 1)}.$$

($T_{jk}(n)$ has one more customer in colony k than n does, hence the appearance of $n_k + 1$ in the arguments.) After substituting and cancelling terms, we see that the partial balance equations are equivalent to

$$\sum_k \lambda_{jk} \stackrel{?}{=} \frac{1}{\alpha_j} \sum_k \alpha_k \lambda_{kj},$$

which is true by the definition of the α_j. □

Remark 2.6 The full balance equations state that the total probability flux into and out of any state is the same. The detailed balance equations state that the total probability flux between any pair of states is the same. Partial balance says that, for a fixed state, there is a subset of the states for which the total probability flux into and out of the subset is equal.

Example 2.7 A telephone banking facility has N incoming lines and a single (human) operator. Calls to the facility are initiated as a Poisson process of rate ν, but calls initiated when all N lines are in use are lost. A call finding a free line has to wait for the operator to answer. The operator deals with waiting calls one at a time, and takes an exponentially distributed length of time with parameter λ to check the caller's identity, after which the call is passed to an automated handling system for the caller to transact banking business, and the operator becomes free to deal with another caller. The automated handling system is able to serve up to N callers simultaneously, and the time it takes to serve a call is exponentially distributed with parameter μ. All these lengths of time are independent of each other and of the Poisson arrival process.

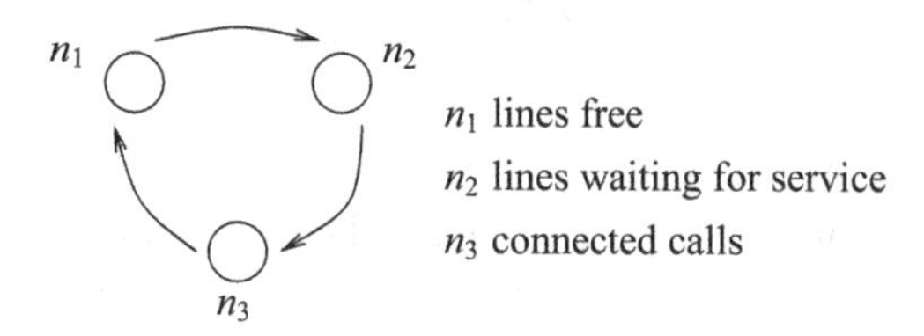

Figure 2.4 Closed migration process for the telephone banking facility.

We model this system as a closed migration process, as in Figure 2.4. The transition rates correspond to

$$\begin{aligned} \lambda_{12} &= \nu, & \phi_1(n_1) &= I[n_1 > 0], \\ \lambda_{23} &= \lambda, & \phi_2(n_2) &= I[n_2 > 0], \\ \lambda_{31} &= \mu, & \phi_3(n_3) &= n_3. \end{aligned}$$

We can easily solve the traffic equations

$$\alpha_1 : \alpha_2 : \alpha_3 = \frac{1}{\nu} : \frac{1}{\lambda} : \frac{1}{\mu}$$

because we have a random walk on three vertices, and these are the average amounts of time it spends in each of the vertices. Therefore, by Theorem 2.4,

$$\pi(n_1, n_2, n_3) \propto \frac{1}{\nu^{n_1}} \frac{1}{\lambda^{n_2}} \frac{1}{\mu^{n_3}} \frac{1}{n_3!}.$$

For example, the proportion of incoming calls that are lost is, by the PASTA property,

$$\mathbb{P}(n_1 = 0) = \sum_{n_2+n_3=N} \pi(0, n_2, n_3).$$

Exercises

Exercise 2.6 For the telephone banking example, show that the proportion of calls lost has the form

$$\frac{H(N)}{\sum_{n=0}^{N} H(n)},$$

where

$$H(n) = \left(\frac{\nu}{\lambda}\right)^n \sum_{i=0}^{n} \left(\frac{\lambda}{\mu}\right)^i \frac{1}{i!}.$$

Exercise 2.7 A restaurant has N tables, with a customer seated at each table. Two waiters are serving them. One of the waiters moves from table to table taking orders for food. The time that he spends taking orders at each table is exponentially distributed with parameter μ_1. He is followed by the wine waiter who spends an exponentially distributed time with parameter μ_2 taking orders at each table. Customers always order food first and then wine, and orders cannot be taken concurrently by both waiters from the same customer. All times taken to order are independent of each other. A customer, after having placed her two orders, completes her meal at rate ν, independently of the other customers. As soon as a customer finishes her meal, she departs, and a new customer takes her place and waits to order. Model this as a closed migration process. Show that the stationary probability that both waiters are busy can be written in the form

$$\frac{G(N-2)}{G(N)} \cdot \frac{\nu^2}{\mu_1 \mu_2}$$

for a function $G(\cdot)$, which may also depend on ν, μ_1, μ_2, to be determined.

2.4 Open migration processes

It is simple to modify the previous model so as to allow customers to enter and exit the system. Define the operators

$$T_{j\to} n = (n_1, \ldots, n_j - 1, \ldots, n_J), \qquad T_{\to k} n = (n_1, \ldots, n_k + 1, \ldots, n_J),$$

where $T_{j\to}$ corresponds to an individual from colony j departing the system; $T_{\to k}$ corresponds to an individual entering colony k from the outside world. We assume that the transition rates associated with these extra possibilities are

$$q(n, T_{jk} n) = \lambda_{jk} \phi_j(n_j); \quad q(n, T_{j\to} n) = \mu_j \phi_j(n_j); \quad q(n, T_{\to k} n) = \nu_k.$$

That is, immigration into colony k is Poisson. We can think of $\rightarrow$ as simply another colony, but with an infinite number of individuals, so that individuals entering or leaving it do not change the overall rate.

If the resulting process n is irreducible in $\mathbb{Z}_+^J$, we call n an *open migration process*.

As before, we define quantities $(\alpha_1, \ldots, \alpha_J)$, which satisfy the new *traffic equations*

$$\alpha_j \left(\mu_j + \sum_k \lambda_{jk} \right) = \nu_j + \sum_k \alpha_k \lambda_{kj}, \qquad \forall j. \tag{2.2}$$

These equations have a unique, positive solution (see Exercise 2.9).

Since the state space of an open migration process is infinite, it is possible that there may not exist an equilibrium distribution. Define the constants

$$g_j = \sum_{n=0}^{\infty} \frac{\alpha_j^n}{\prod_{r=1}^n \phi_j(r)}, \qquad j = 1, 2, \ldots, J.$$

Theorem 2.8 *If $g_1, \ldots, g_J < \infty$, then n has the equilibrium distribution*

$$\pi(n) = \prod_{j=1}^J \pi_j(n_j), \qquad \pi_j(n_j) = g_j^{-1} \frac{\alpha_j^{n_j}}{\prod_{r=1}^{n_j} \phi_j(r)}.$$

Proof Once again, we need to check for each n the full balance equations,

$$\begin{aligned}\pi(n)\left(\sum_j \sum_k q(n, T_{jk}n) + \sum_j q(n, T_{j\rightarrow}n) + \sum_k q(n, T_{\rightarrow k}n) \right)& \\ \stackrel{?}{=} \sum_j \sum_k \pi(T_{jk}n) q(T_{jk}n, n) + \sum_j \pi(T_{j\rightarrow}n) q(T_{j\rightarrow}n, n)& \\ + \sum_k \pi(T_{\rightarrow k}n) q(T_{\rightarrow k}n, n),&\end{aligned}$$

which will be satisfied if we can solve the partial balance equations,

$$\begin{aligned}\pi(n)\left(\sum_k q(n, T_{jk}n) + q(n, T_{j\rightarrow}n) \right)& \\ \stackrel{?}{=} \sum_k \pi(T_{jk}n) q(T_{jk}n, n) + \pi(T_{j\rightarrow}n) q(T_{j\rightarrow}n, n)& \qquad (2.3)\end{aligned}$$

and

$$\pi(n) \sum_k q(n, T_{\rightarrow k}n) \stackrel{?}{=} \sum_k \pi(T_{\rightarrow k}n) q(T_{\rightarrow k}n, n). \tag{2.4}$$

(These look a lot like the earlier partial balance equations with an added colony called $\to$.) Substitution of the claimed form for $\pi(n)$ will verify that equations (2.3) are satisfied (check this!).

To see that equations (2.4) are satisfied, we substitute

$$q(n, T_{\to k}n) = \nu_k, \quad \pi(T_{\to k}n) = \pi(n)\frac{\alpha_k}{\phi_k(n_k+1)}, \quad q(T_{\to k}n, n) = \mu_k\phi_k(n_k+1),$$

to obtain, after cancellation,

$$\sum_k \nu_k \stackrel{?}{=} \sum_k \alpha_k\mu_k.$$

This is not directly one of the traffic equations (2.2): instead, it is the sum of these equations over all j. □

We conclude that, in equilibrium, at any fixed time t the random variables $n_1(t), n_2(t), \ldots, n_J(t)$ are independent (although in most open migration networks they won't be independent as processes).

A remarkable property of the distribution $\pi_j(n_j)$ is that it is the same as if the arrivals at colony j were a Poisson process of rate $\alpha_j\lambda_j$, with departures happening at rate $\lambda_j\phi_j(n_j)$, where we define $\lambda_j = \mu_j + \sum_k \lambda_{jk}$. However, in general the entire process of arrivals of individuals into a colony is not Poisson. For example, consider the system in Figure 2.5. If ν is quite small but λ_{21}/μ_2 is quite large, the typical arrival process into queue 1 will look like the right-hand picture: rare bursts of arrivals of geometric size.

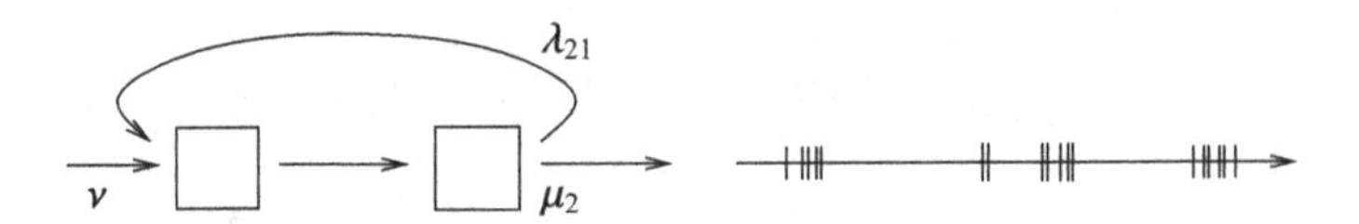

Figure 2.5 Simple open migration network. The typical arrival process into queue 1 is illustrated on the right.

While an open migration process is not in general reversible (the detailed balance equations do not hold), it does behave nicely under time reversal: the reversed process also looks like an open migration process.

Theorem 2.9 *If $(n(t), t \in \mathbb{R})$ is a stationary open migration process, then so is the reversed process $(n(-t), t \in \mathbb{R})$.*

Proof We need to check that the transition rates have the required form.

Using Proposition 1.1, we have

$$q'(n, T_{jk}n) = \frac{\pi(T_{jk}n)}{\pi(n)} q(T_{jk}n, n) = \frac{\phi_j(n_j)}{\alpha_j} \frac{\alpha_k}{\phi_k(n_k+1)} \lambda_{kj} \phi_k(n_k+1)$$

$$= \lambda'_{jk}\phi_j(n_j), \quad \text{where} \quad \lambda'_{jk} = \frac{\alpha_k}{\alpha_j}\lambda_{kj}.$$

Similarly we find (check this) that the remaining transition rates of the reversed open migration process have the form

$$q'(n, T_{j\to}n) = \mu'_j \phi_j(n_j), \quad q'(n, T_{\to k}n) = \nu'_k,$$

where $\mu'_j = \nu_j/\alpha_j$ and $\nu'_k = \alpha_k \mu_k$. □

Corollary 2.10 *The exit process from colony k, i.e. the stream of individuals leaving the system from colony k, is a Poisson process of rate* $\alpha_k \mu_k$.

Of course, as Figure 2.5 shows, not all of the internal streams of individuals can be Poisson! However, more of them may be Poisson than has been asserted so far.

Remark 2.11 Consider the open migration process illustrated in Figure 2.6; the circles are colonies and the arrows indicate positive values of λ, ν, μ.

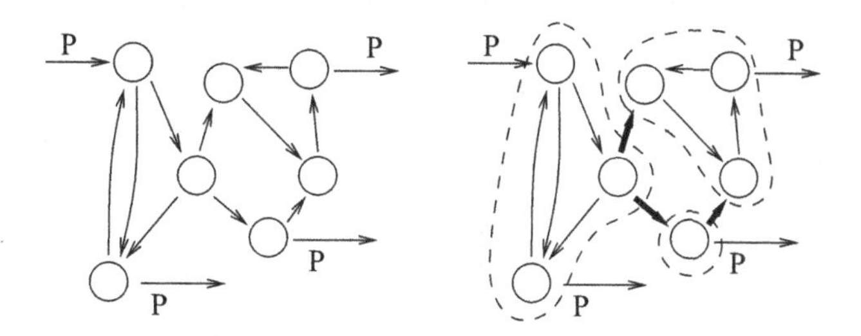

Figure 2.6 Example open migration network. The dashed sets of colonies will be useful in Exercise 2.10.

In Exercise 2.10 we will show that the stream of customers is Poisson, not only along those arrows marked with a P, but also along all the thicker arrows.

We have so far tacitly assumed that the open migration process has a finite number of colonies, but we haven't explicitly used this assumption; the only extra condition we require to accommodate $J = \infty$ is to insist that the equilibrium distribution can be normalized, i.e. that $\prod_{j=1}^{\infty} g_j^{-1} > 0$. We will now discuss an example of an open migration process with J infinite.

Family-size process

This process was studied by Kendall (1975), who was interested in family names in Yorkshire. An immigrating individual has a distinguishing characteristic, such as a surname or a genetic type, which is passed on to all his descendants. The population is divided into families, each of which consists of all those individuals alive with a given characteristic. The process can also be used to study models of preferential attachment, where new objects (e.g. web pages or news stories) tend to attach to popular objects.

Let n_j be the number of families of size j. Then the family-size process $(n_1, n_2, \dots)$ is a linear open migration process with transition rates

$$
\begin{aligned}
q(n, T_{j,j+1}n) &= j\lambda n_j, & j &= 1, 2, \dots, && - \lambda \text{ is the birth rate}\\
q(n, T_{j,j-1}n) &= j\mu n_j, & j &= 2, 3, \dots, && - \mu \text{ is the death rate}\\
q(n, T_{\to 1}n) &= \nu, & & && - \text{immigration}\\
q(n, T_{1\to}n) &= \mu n_1, & & && - \text{extinction of a family.}
\end{aligned}
$$

This corresponds to an open migration process with $\phi_j(n_j) = n_j$ (and, for example, $\lambda_{j,j+1} = j\lambda$). Observe that a *family* is the basic unit which moves through the colonies of the system, and that the movements of different families are independent.

The traffic equations have a solution (check!)

$$\alpha_j = \frac{\nu}{\lambda j}\left(\frac{\lambda}{\mu}\right)^j,$$

with

$$g_j = \sum_{n=0}^{\infty} \frac{\alpha_j^n}{n!} = e^{\alpha_j}, \qquad \prod_{i=1}^{\infty} g_i^{-1} = e^{-\sum \alpha_i} > 0 \text{ if } \lambda < \mu.$$

The condition $\lambda < \mu$ is to be expected, for stability. Under this condition, the equilibrium distribution is

$$\pi(n) = \prod_{j=1}^{\infty} e^{-\alpha_j} \frac{\alpha_j^{n_j}}{n_j!},$$

and thus the number of families of size j, n_j, has a Poisson distribution with mean α_j, and $n_1, n_2, \dots$ are independent. Hence the total number of distinct families, $N = \sum_j n_j$, has a Poisson distribution with mean

$$\sum_j \alpha_j = -\frac{\nu}{\lambda} \log\left(1 - \frac{\lambda}{\mu}\right),$$

from the series expansion of $\log(1 - x)$.

Exercises

Exercise 2.8 Show that the reversed process obtained from a stationary closed migration process is also a closed migration process, and determine its transition rates.

Exercise 2.9 Show that there exists a unique, positive solution to the traffic equations (2.2).
[*Hint:* Define $\alpha_{\rightarrow} = 1$.]

Exercise 2.10 Consider Figure 2.6. Show that each of the indicated subsets of colonies is itself an open migration process.
[*Hint:* Induction.]

Suppose that, in an open migration process, individuals from colony k cannot reach colony j without leaving the system. Show that the process of individuals transitioning directly from colony j to colony k is Poisson. Conclude that the processes of individuals transitioning along bold arrows are Poisson.
[*Hint:* Write $j \rightsquigarrow k$ if individuals leaving colony j can reach colony k by some path with positive probability. Form subsets of colonies as equivalence classes, using the equivalence relation: $j \sim k$ if both $j \rightsquigarrow k$ and $k \rightsquigarrow j$.]

Exercise 2.11 In the family-size process show that the points in time at which family extinctions occur form a Poisson process.

Exercise 2.12 In the family-size process, let

$$M = \sum_{1}^{\infty} j n_j$$

be the population size. Determine the distribution of n conditional on M, and show that it can be written in the form

$$\pi(n \mid M) = \binom{\theta + M - 1}{M}^{-1} \prod_{j=1}^{M} \left(\frac{\theta}{j}\right)^{n_j} \frac{1}{n_j!}, \tag{2.5}$$

where $\theta = \nu/\lambda$.
[*Hint:* In the power series expansions of the identity

$$(1 - x)^{-\theta} = \prod_{j=1}^{\infty} \exp\left(\frac{\theta}{j} x^j\right),$$

consider the coefficients of x^M.]

Exercise 2.13 (The Chinese restaurant process) An initially empty restaurant has an unlimited number of tables, each capable of seating an unlimited number of customers. Customers numbered $1, 2, \ldots$ arrive one by one and are seated as follows. Customer 1 sits at a new table. For $m \geq 1$ suppose m customers are seated in some arrangement: then customer $m+1$ sits at a new table with probability $\theta/(\theta + m)$ and sits at a given table already seating j customers with probability $j/(\theta + m)$.

Let n_j be the number of tables occupied by j customers. Show that after the first M customers have arrived the distribution of $(n_1, n_2, \ldots)$ is given by expression (2.5).

Deduce that for the family-size process the expected number of distinct families (e.g. surnames, genetic types, or news stories) given the population size is

$$\mathbb{E}(N \mid M) = \sum_{i=1}^{M} \frac{\theta}{\theta + i - 1}.$$

2.5 Little's law

Consider a stochastic process $(X(t), t \geq 0)$ on which is defined a function $n(t) = n(X(t))$, the *number of customers in the system* at time t. Suppose that with probability 1 there exists a time T_1 such that the continuation of the process beyond T_1 is a probabilistic replica of the process starting at time 0: this implies the existence of further times $T_2, T_3, \ldots$ having the same property as T_1. These times are called *regeneration points*. Suppose also that the regeneration point is such that $n(T_1) = 0$. (For example, in an open migration process a regeneration point may be "entire system is empty".) Suppose $\mathbb{E}T_1 < \infty$. Then we can define the mean number of customers in the system as follows:

$$\lim_{T \to \infty} \frac{1}{T} \int_0^T n(s)ds \overset{w.p.1}{=} \lim_{T \to \infty} \frac{1}{T} \int_0^T \mathbb{E}n(s)ds = \frac{\mathbb{E} \int_0^{T_1} n(s)ds}{\mathbb{E}T_1} \equiv L$$

(that is, the average number of customers in the system can be computed from a single regeneration cycle – treat this as an assertion for the moment).

Also, let $W_n, n = 1, 2, \ldots,$ be the amount of time spent by the nth customer in the system. Then we can define the mean time spent by a customer in the system as follows:

$$\lim_{n \to \infty} \frac{1}{n} \sum_{i=1}^{n} W_n \overset{w.p.1}{=} \lim_{n \to \infty} \frac{1}{n} \sum_{i=1}^{n} \mathbb{E}W_n = \frac{\mathbb{E} \sum_{n=1}^{N} W_n}{\mathbb{E}N} \equiv W,$$

where N is the number of customers who arrive during the first regeneration cycle $[0, T_1]$.

Finally, we can define the mean arrival rate

$$\lim_{T\to\infty} \frac{1}{T}(\text{number of arrivals in } [0,T])$$

$$\stackrel{w.p.1}{=} \lim_{T\to\infty} \frac{1}{T}\mathbb{E}(\text{number of arrivals in } [0,T]) = \frac{\mathbb{E}N}{\mathbb{E}T_1} \equiv \lambda.$$

Note that we do not assume that the arrivals are Poisson; λ is simply the long-run average arrival rate.

Remark 2.12 We will not prove any of these statements. (They follow from renewal theory, and are discussed further in Appendix B.) But we will establish a straightforward consequence of these statements.

Note that the definition of what constitutes the system, or a customer, is left flexible. We require only consistency of meaning in the phrases "number of customers in the system", "time spent in the system", "number of arrivals". For example, suppose that in an open migration process we want to define "the system" to be a subset of the colonies. Then $n(t)$ is the number of customers present in the subset of the colonies, but we have a choice for the arrivals and time periods. We could view each visit of an individual to the subset as a distinct arrival, generating a distinct time spent in "the system". Alternatively, we could count each distinct individual as a single arrival, and add up the times of that individual's visits into a single time spent in the system.

Theorem 2.13 (Little's law) $L = \lambda W$.

Proof Note

$$L = \frac{\mathbb{E}\int_0^{T_1} n(s)ds}{\mathbb{E}T_1} = \frac{\mathbb{E}\int_0^{T_1} n(s)ds}{\mathbb{E}N}\frac{\mathbb{E}N}{\mathbb{E}T_1},$$

whereas

$$\lambda W = \frac{\mathbb{E}N}{\mathbb{E}T_1}\frac{\mathbb{E}\sum_{i=1}^N W_i}{\mathbb{E}N}.$$

Thus, it suffices to show

$$\int_0^{T_1} n(s)ds = \sum_{i=1}^N W_i. \tag{2.6}$$

Figure 2.7 illustrates this equality for a simple queue: the shaded area can be calculated as an integral over time, or as a sum over customers. More generally, we see equality holds by the following argument. Imagine each customer pays at a rate of 1 per unit time while in the system (and so the total paid by the nth customer in the system is just W_n). Then the two sides of (2.6) are just different ways of calculating the total amount paid during a regenerative cycle.

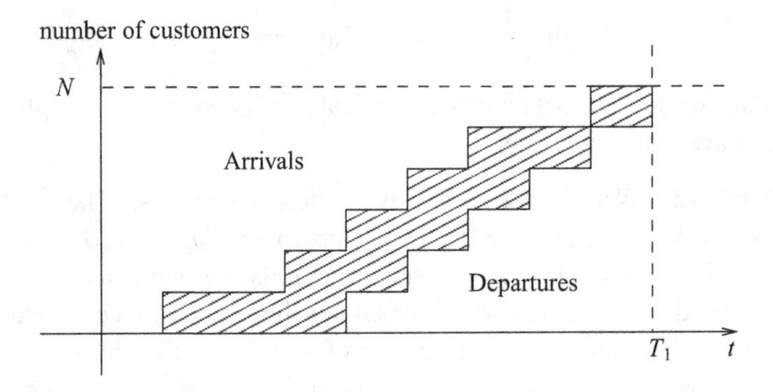

Figure 2.7 Shaded area is $\int_0^{T_1} n(s)ds$, but also $\sum_{i=1}^N W_i$.

□

Remark 2.14 Little's law holds whenever we can reasonably define the quantities L, λ and W. The proof of the relationship between them is going to be essentially the same; the question is simply whether it makes sense to talk of the average arrival rate, the number of customers in the system, or the time spent in the system.

Exercises

Exercise 2.14 The Orchard tea garden remains open 24 hours per day, 365 days per year. The total number of customers served in the tea garden during 2012 was 21% greater than the total for 2011. In each year, the number of customers in the tea garden was recorded at a large number of randomly selected times, and the average of those numbers in 2012 was 16% greater than the average in 2011. By how much did the average duration of a customer visit to the Orchard increase or decrease?

Exercise 2.15 Use Little's law to check the result in Exercise 1.7.

Exercise 2.16 Customers arrive according to a Poisson process with rate ν at a single server, but a restricted waiting room causes those who arrive when n customers are already present to be lost. Accepted customers have service times which are independent and identically distributed with mean μ, and independent of the arrival process. (Service times are not necessarily exponentially distributed.) Identify a regeneration point for the system. If P_j is the long-run proportion of time j customers are present, show that

$$1 - P_0 = \nu\mu(1 - P_n).$$

Exercise 2.17 Consider two possible designs for an S-server queue. We may have all the customers wait in a single queue; the person at the head of the line will be served by the first available server. (This is the arrangement at the Cambridge railway ticket office.) Alternatively, we might have S queues (one per server), with each customer choosing a queue to join when she enters the system. Assuming that customers can switch queues, and in particular will not allow a server to go idle while anyone is waiting, is there a difference in the expected waiting time between the two systems? Is there a difference in the variance?

2.6 Linear migration processes

In this section we consider systems which have the property that after individuals (or customers or particles) have entered the system they move independently through it. We have seen the example of linear open migration networks, where in equilibrium the numbers in distinct colonies are independent Poisson random variables. We will see that this result can be be substantially generalized, to time-dependent linear migration processes on a general space. We'll need a little more mathematical infrastructure, which is useful also to gain a better understanding of Poisson processes.

Let $\mathcal{X}$ be a set, and $\mathcal{F}$ a σ-algebra of measurable subsets of $\mathcal{X}$. (If you are not familiar with measure theory, then just view the elements of $\mathcal{F}$ as the subsets of $\mathcal{X}$ in which we are interested.)

A *Poisson point process with mean measure* M is a point process for which, if $E_1, \dots, E_k \in \mathcal{F}$ are disjoint sets, the number of points in $E_1, \dots, E_k \in \mathcal{F}$ are independent Poisson random variables, with $M(E_i)$ the mean number of points in E_i.

As a simple example, let $\mathcal{X} = \mathbb{R}$, let $\mathcal{F}$ be the Borel σ-algebra generated by open intervals, and let M be Lebesgue measure (intuitively, $M(E)$ measures the size of the set E – its length if E is an interval – and $E \in \mathcal{F}$ if the size can be sensibly defined). The probability of no points in the interval $(0, t)$ is, from the Poisson distribution, e^{-t}; hence the distance from 0 to the first point to the right of 0 has an exponential distribution. Similarly, the various other properties of a Poisson process indexed by time can be deduced. For example, given there are N points in the interval $(0, t)$, their positions are independent and uniformly distributed.

Another example is a Poisson process on $[0, 1] \times [0, \infty)$, with the Borel σ-algebra and Lebesgue measure. We can think of $[0, 1]$ as "space" and $[0, \infty)$ as "time": then this is a sequence of points arriving as a unit rate Poisson process in time, and each arriving point is distributed uniformly on $[0, 1]$, independently of all the other points. The collections of points arriving to different subsets of $[0, 1]$, or over different time periods, are independent. If we colour red those points whose space coordinate lies in $(0, p)$, and blue those points whose space coordinate lies in $(p, 1)$, for $0 < p < 1$, then the times of arrival of red and blue points form independent Poisson processes with rates p and $1 - p$, respectively. The arrival process of red points is formed by *thinning* the original arrival process: an arriving point is retained with probability p, independently from point to point. Similarly, if two independent Poisson processes with mean measures M_1, M_2 are superimposed, we obtain a Poisson process with mean measure $M_1 + M_2$.

Suppose now that individuals arrive into $\mathcal{X}$ as a Poisson stream of rate ν, and then proceed to move independently through $\mathcal{X}$ according to some stochastic process before (possibly) leaving $\mathcal{X}$. We would like to understand the collection of points that represents the (random) set of individuals in the system at any given time.

Example 2.15 Recall our earlier discussion of linear migration processes, in Section 2.4. Then we can take $\mathcal{X}$ to be the set of colonies $\{1, 2, \ldots, J\}$ or $\{1, 2, \ldots\}$. We assume that individuals enter the system into colony j as a Poisson process of rate ν_j. After arriving at colony j, an individual stays there for a certain random amount of time T_j with mean λ_j^{-1}, the *holding time* of the individual in colony j. At the end of its holding time in colony j, an individual moves to colony k with probability $p(j, k)$ or leaves the system with probability $p(j, \rightarrow)$.

In our earlier discussion of linear migration processes in Section 2.4, the times T_j were exponentially distributed with parameter λ_j, and all holding times were independent. But now, in this example, we relax this assumption, allowing T_j to be arbitrarily distributed with mean λ_j^{-1}, and we allow the holding times of an individual at the colonies she visits to be dependent. (But we insist these holding times are independent of those of other individuals, and of the Poisson arrival process.) In Exercise 2.20, we will show that this does not change the limiting distribution for the numbers in different colonies, a result known as *insensitivity*.

Example 2.16 Consider cars moving on a highway. The set $\mathcal{X} = \mathbb{R}_+$ corresponds to a semi-infinite road. Cars arrive at the point 0 as a Poisson process, and then move forward (i.e. right). If the highway is wide and uncongested, we may assume that they move independently of each other, passing if necessary. At any time, the set of car locations is a random collection of points in $\mathcal{X}$, and we are interested in properties of this random collection as it evolves in time. To determine the stochastic process fully, we need to specify the initial state of the highway at some time $t = 0$, and the way in which each car moves through the system. We will do this in Exercise 2.18.

For $E \in \mathcal{F}$ define the quantity

$$P(u, E) = \mathbb{P}(\text{individual is in } E \text{ a time } u \text{ after her arrival into the system}).$$

Theorem 2.17 (Bartlett's theorem) *If the system is empty at time* 0 *then at time t individuals are distributed over $\mathcal{X}$ according to a Poisson process with mean measure*

$$M(t, E) = \nu \int_0^t P(u, E)\, du, \qquad E \in \mathcal{F}.$$

Remark 2.18 The form for the mean measure should not be a surprise: the mean number of individuals in E at time t is obtained by integrating over $(0, t)$ the rate at which individuals arrive multiplied by the probability that at time t they are in E. But we need to prove that the numbers in disjoint subsets are independent, with a Poisson distribution.

Proof Let $E_1, \dots, E_k \in \mathcal{F}$ be disjoint, and let $n_j(t)$ be the number of individuals in E_j at time t. In order to calculate the joint distribution of the random variables $n_j(t)$, we will compute the joint probability generating function. We will use the shorthand $z^{n(t)}$ to mean $z_1^{n_1(t)} \dots z_k^{n_k(t)}$; the probability generating function we want is then $\mathbb{E}z^{n(t)}$.

Let m be the number of arrivals into the system in $(0, t)$. Conditional on m, the arrival times $\tau_1, \dots, \tau_m$ are independent and uniform in $(0, t)$ (unordered, of course). Let A_{ri} be the event that the individual who arrived at time τ_r is in E_i at time t; then $\mathbb{P}(A_{ri}) = P(t - \tau_r, E_i)$.

Now,

$$\begin{aligned}
\mathbb{E}[z^{n(t)} \mid m, \tau_1, \dots, \tau_m] &= \mathbb{E}\left[\prod_{r=1}^{m}\prod_{i=1}^{k} z_i^{I[A_{ri}]} \mid m, \tau_1, \dots, \tau_m\right] \\
&= \prod_{r=1}^{m} \mathbb{E}\left[\prod_{i=1}^{k} z_i^{I[A_{ri}]} \mid \tau_r\right] \quad \text{(by independence of } \tau_r) \\
&= \prod_{r=1}^{m}\left(1 - \sum_{i=1}^{k}(1 - z_i)P(t - \tau_r, E_i)\right),
\end{aligned}$$

by considering the various places the arrival at time τ_r could be (in at most one of E_i, $i = 1, 2, \dots, k$). Taking the average over τ_r, we get

$$\begin{aligned}
\mathbb{E}[z^{n(t)} \mid m] &= \prod_{r=1}^{m}\left(1 - \sum_{i=1}^{k}(1 - z_i)\frac{1}{t}\int_0^t P(t - \tau, E_i)\, d\tau\right) \\
&= \left(1 - \sum_{i=1}^{k}(1 - z_i)\frac{1}{t}\int_0^t P(t - \tau, E_i)\, d\tau\right)^m.
\end{aligned}$$

To take the average over m, note that m is a Poisson random variable with mean νt, and for a Poisson random variable X with mean λ we have $\mathbb{E}z^X = e^{-(1-z)\lambda}$. Therefore,

$$\begin{aligned}
\mathbb{E}[z^{n(t)}] &= \exp\left(-\nu t \sum_{i=1}^{k}(1 - z_i)\frac{1}{t}\int_0^t P(u, E_i)\, du\right) \\
&= \prod_{i=1}^{k} \exp(-(1 - z_i)M(t, E_i)).
\end{aligned}$$

This shows that the joint probability generating function of $n_1(t), \dots, n_k(t)$ is the product of the probability generating functions of Poisson random variables with means $M(t, E_i)$; therefore, $n_1(t), \dots, n_k(t)$ are independent Poisson random variables with the claimed means, as required. □

Remark 2.19 Some intuition (and a rather slicker proof) for the above result can be developed as follows. Colour an individual who arrives at the system at time τ with colour j if that individual is in E_j at time t, for $j = 1, \dots, J$, and leave it uncoloured if it is not in any of these sets at time t. Then the colours of individuals are independent, and the probability an arrival at time τ is coloured j is $P(t-\tau, E_j)$. We are interested in the colours of the arrivals over $(0, t)$, and the result then follows from general thinning and superposition results for Poisson processes.

Corollary 2.20 *As $t \to \infty$ the distribution of individuals over $\mathcal{X}$ approaches a Poisson process with mean measure*

$$M(E) = \nu \int_0^\infty P(u, E)\, du, \quad E \in \mathcal{F}.$$

Remark 2.21 Observe that $\int_0^\infty P(u, E)\, du$ is the mean time spent by an individual in the set E, and so this expression for the mean number in the set E is Little's law. (Of course, Little's law does not give the distribution, only the mean.)

Exercises

Exercise 2.18 Cars arrive at the beginning of a long stretch of road in a Poisson stream of rate ν from time $t = 0$ onwards. A car has a fixed velocity $V > 0$, which is a random variable. The velocities of cars are independent and identically distributed, and independent of the arrival process. Cars can overtake each other freely. Show that the number of cars on the first x miles of road at time t has a Poisson distribution with mean

$$\nu \mathbb{E}\left[\frac{x}{V} \wedge t\right].$$

Exercise 2.19 Airline passengers arrive at a passport control desk in accordance with a Poisson process of rate ν. The desk operates as a single-server queue at which service times are independent and exponentially distributed with mean $\mu(< \nu^{-1})$ and are independent of the arrival process. After leaving the passport control desk, a passenger must pass through a security check. This also operates as a single-server queue, but one at which service times are all equal to $\tau(< \nu^{-1})$. Show that in equilibrium the probability both queues are empty is

$$(1 - \nu\mu)(1 - \nu\tau).$$

If it takes a time σ to walk from the first queue to the second, what is the equilibrium probability that both queues are empty and there is no passenger walking between them?
[*Hint:* Use Little's law to find the probability the second queue is empty; use the approach of Section 2.2 to establish independence; and use the properties of a Poisson process to determine the distribution of the number of passengers walking between the two queues.]

Exercise 2.20 Consider the linear migration process of Example 2.15.

Write down the limiting distribution of the vector $(n_1, n_2, \ldots, n_J)$, the numbers of individuals in the J colonies. Show that the limiting distribution of n_j depends on the distribution of T_j only through $\mathbb{E}T_j$, the expected amount of time that an individual customer spends in colony j. The limiting distribution is *insensitive* to the precise form of the holding time distribution, as well as to dependencies between an individual's holding times at colonies visited.

2.7 Generalizations

The elementary migration processes considered earlier in this chapter can be generalized in various ways, while still retaining the simplicity of a product-form equilibrium distribution. We shall not go into this in great detail, but with a few examples we briefly sketch two generalizations, namely multiple classes of customer and general service distributions.

Example 2.22 (Communication network/optimal allocation) Our first example is of a network (e.g., of cities) with one-way communication links between them. The links are used to transmit data packets; and if there are capacity constraints on the links, it is possible that packets may have to queue for service. If we model the links as queues (as in the right-hand

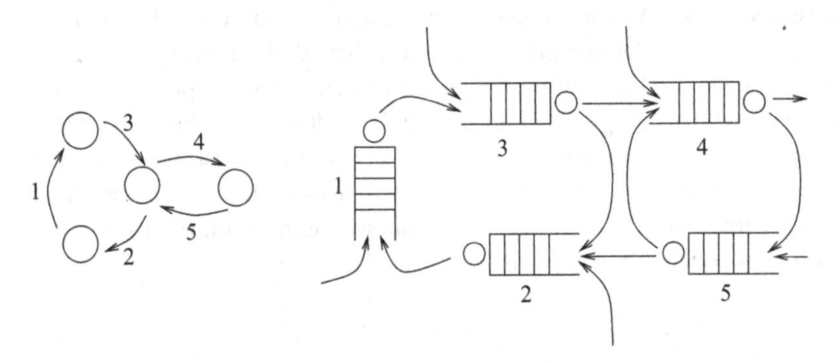

Figure 2.8 A communication network. On the right is shown the same network with links modelled as queues.

diagram of Figure 2.8), we get something that isn't quite an open migration network; in an open migration network, individuals can travel in cycles, whereas packets would not do that. (In fact, a natural model of this system is as a *multi-class* queueing network, but we will not be developing that framework here.)

Let a_j be the average arrival rate of messages at queue j, and let n_j be the number of messages in queue j. Suppose that we can establish that in equilibrium

$$\mathbb{P}(n_j = n) = \left(1 - \frac{a_j}{\phi_j}\right)\left(\frac{a_j}{\phi_j}\right)^n,$$

where ϕ_j is the service rate at queue j. (It is possible to prove under some assumptions that the equilibrium distribution of a multi-class queueing network is of product form, with these distributions as marginal distributions.)

Suppose now that we are allowed to choose the service rates (capacities) $\phi_1, \ldots, \phi_J$, subject only to the budget constraint $\sum \phi_j = F$, where all summations are over $j = 1, 2, \ldots, J$. What choice will minimize the mean number of packets in the system, or (equivalently, in view of Little's law) the mean period spent in the system by a message?

The optimization problem we are interested in solving is

$$\begin{aligned} &\text{minimize } \mathbb{E}\left(\sum n_j\right) = \sum \frac{a_j}{\phi_j - a_j}, \\ &\text{subject to } \sum \phi_j = F, \\ &\text{over} \qquad \phi_j \geq 0, \qquad \forall j. \end{aligned}$$

We shall solve this problem using Lagrangian techniques (see Appendix C for a review). Introduce a Lagrange multiplier y for the budget constraint, and let

$$L(\phi; y) = \sum \frac{a_j}{\phi_j - a_j} + y\left(\sum \phi_j - F\right).$$

Setting $\partial L / \partial \phi_j = 0$, we find that L is minimized over ϕ by the choice

$$\phi_j = a_j + \sqrt{a_j / y}.$$

Substitution of this into the budget constraint and solving for y shows that we should choose

$$\phi_j = a_j + \frac{\sqrt{a_j}}{\sum \sqrt{a_k}}\left(F - \sum a_k\right).$$

This is known as *Kleinrock's square root channel capacity assignment*: any excess capacity (beyond the bare minimum required to service the mean arrival rates) is allocated proportionally to the square root of the arrival rates. This approach was used by Kleinrock (1964, 1976) for the early computer network ARPANET, the forerunner of today's Internet, when capacity was expensive and queueing delays were often lengthy.

Example 2.23 (Processor sharing) Consider a model in which K individuals, or jobs, oscillate between being served by a processor and being idle, where the processor has a total service rate μ that it shares out equally among all the n_0 active jobs. (One might imagine K customers using a web server for their shopping, with idleness corresponding to waiting for a customer's response.) Each job has an exponential service requirement of unit mean at the processor, queue 0, and so if n_0 jobs are at queue 0 then each one of them departs at rate μ/n_0. Suppose when job k becomes idle it remains so for an exponentially distributed time with parameter λ_k. The schematic diagram appears in Figure 2.9.

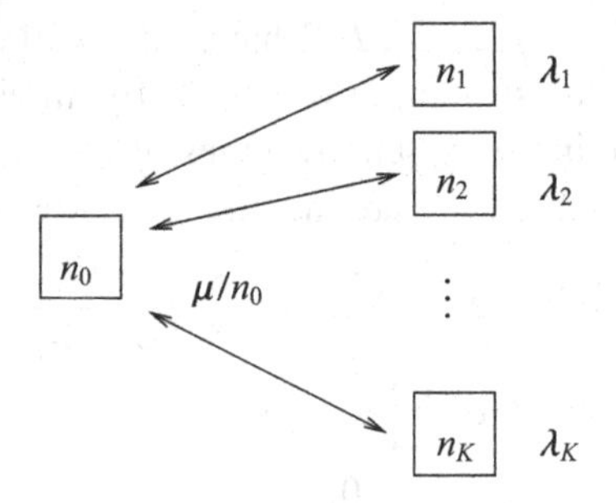

Figure 2.9 Processor-sharing system: n_0 jobs are currently being served, and $n_k = 0$ or 1 for $k \neq 0$.

In Exercise 2.21, you will check that the stationary distribution for this system is

$$\pi(n_0, \ldots, n_K) \propto n_0! \mu^{-n_0} \prod_{k=1}^{K} \lambda_k^{-n_k} \tag{2.7}$$

and that it satisfies detailed balance.

The interesting fact about this example is that while we need quite specific distributional assumptions (exponential service requirements and idle times) in order to check detailed balance, the stationary distribution for $(n_0, \ldots, n_K)$ is actually the same if the distributions are quite arbitrary with the same mean. There are several scheduling disciplines for which this sort of *insensitivity* holds, including the queues of this example. The Erlang model of Section 1.3 is insensitive to the distribution of call holding periods, as is its network generalization in Chapter 3. We looked briefly at this phenomenon in Exercise 2.20, but we shall not study it in more detail. (Notably, a single-server queue with first-come-first-served scheduling discipline *is* sensitive to its service time distribution.)

Example 2.24 (A wireless network) Our next example is a model of a network of K nodes sharing a wireless medium. Let the nodes form the vertex set of a graph, and place an edge between two nodes if the two nodes interfere with each other. Call the resulting undirected graph the *interference graph* of the network. Let $\mathcal{S} \subset \{0,1\}^K$ be the set of vectors $n = (n_1, n_2, \ldots, n_K)$ with the property that if i and j have an edge between them then $n_i \cdot n_j = 0$. Let n be a Markov process with state space $\mathcal{S}$ and transition rates

$$\begin{aligned} q(n, n - e_k) &= \mu_k \qquad \text{if} \qquad n_k = 1, \\ q(n, n + e_k) &= \nu_k \qquad \text{if} \qquad n + e_k \in \mathcal{S}, \end{aligned}$$

where $e_k \in \mathcal{S}$ is a unit vector with a 1 as its kth component and 0s elsewhere. We interpret $n_k = 1$ as indicating that node k is transmitting. Thus a node is blocked from transmitting whenever any of its neighbours are active. Node k starts transmitting at rate ν_k provided it is not blocked, and transmits for an exponentially distributed period of time with parameter μ_k.

It is immediate that the stationary distribution for this system is

$$\pi(n_1, \ldots, n_K) \propto \prod_{k=1}^{K} \left(\frac{\nu_k}{\mu_k} \right)^{n_k} \tag{2.8}$$

normalized over the state space $\mathcal{S}$, and that it satisfies detailed balance. (Indeed, the distribution is unaltered if the periods for which node k transmits are independent random variables arbitrarily distributed with mean $1/\mu_k$, and if the periods for which node k waits before attempting a transmission following either the successful completion of, or a blocked attempt at, a transmission are independent random variables arbitrarily distributed with mean $1/\nu_k$.)

We look at variants of this model in Chapters 3, 5 and 7.

Exercises

Exercise 2.21 Check (by verifying detailed balance) that expression (2.7) gives the stationary distribution for the processor-sharing example.

Exercise 2.22 Deduce, from the form (2.8), that if $\lambda = \nu_k/\mu_k$ does not depend on k then the stationary probability of any given configuration $(n_1, \ldots, n_K)$ depends on the number of active transmissions, $\sum_k n_k$, but not otherwise on which stations are transmitting.

An *independent vertex set* is a set of vertices of the interference graph with the property that no two vertices are joined by an edge. Show that as

$\lambda \to \infty$ the stationary probability approaches a uniform distribution over the independent vertex sets of maximum cardinality.

2.8 Further reading

Whittle (1986), Walrand (1988), Bramson (2006) and Kelly (2011) give more extended treatments of migration processes and queueing networks, including the generalizations of Section 2.7. Pitman (2006) provides an extensive treatment of the Chinese restaurant process (Exercise 2.13). Baccelli and Brémaud (2003) develop a very general framework for the study of queueing results such as Little's law and the PASTA property. Kingman (1993) is an elegant text on Poisson processes on general spaces.

The short-hand queueing notation was suggested by Kendall (1953): thus a $D/G/K$ queue has a deterministic arrival process, general service times, and K servers. Asmussen (2003) provides an authoritative treatment of queues, and of the basic mathematical tools for their analysis.

3

Loss networks

We have seen, in Section 1.3, Erlang's model of a telephone link. But what happens if the system consists of many links, and if calls of different types (perhaps voice, video or conference calls) require different resources? In this chapter we describe a generalization of Erlang's model, which treats a network of links and which allows the number of circuits required to depend upon the call. The classical example of this model is a telephone network, and it is natural to couch its definition in terms of calls, links and circuits. Circuits may be physical (in Erlang's time, a strand of copper; or, for a wireless link, a radio frequency) or virtual (a fixed proportion of the transmission capacity of a communication link such as an optical fibre). The term "circuit-switched" is common in some application areas, where it is used to describe systems in which, before a request (which may be a call, a task or a customer) is accepted, it is first checked that sufficient resources are available to deal with each stage of the request. The essential features of the model are that a call makes simultaneous use of a number of resources and that blocked calls are lost.

The simplest model of a loss network is a single link with C circuits, where the arrivals are Poisson of rate ν, the holding times are exponential with mean 1, and blocked calls are lost. In this case, we know from Erlang's formula (1.5)

$$\mathbb{P}(\text{an arriving call is lost}) = E(\nu, C) = \frac{\nu^C}{C!}\left(\sum_{j=0}^{C}\frac{\nu^j}{j!}\right)^{-1}.$$

3.1 Network model

Let the set of links be $\mathcal{J} = \{1, 2, \ldots, J\}$, and let C_j be the number of circuits on link j. A *route* r is a subset of $\{1, 2, \ldots, J\}$, and the set of all routes is called $\mathcal{R}$ (this is a subset of the power set of $\mathcal{J}$). Let $R = |\mathcal{R}|$. Figure 3.1

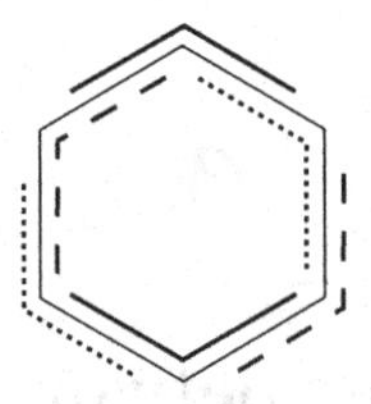

Figure 3.1 A loss network with six links and six two-link routes.

shows an example of a network with six links, in which "routes" are pairs of adjacent links.

Calls requesting route r arrive as a Poisson process of rate ν_r, and as r varies it indexes independent Poisson streams. A call requesting route r is blocked and lost if on any link $j \in r$ there is no free circuit. Otherwise the call is connected and simultaneously holds one circuit on each link $j \in r$ for the holding period of the call. The call holding period is exponentially distributed with unit mean and independent of earlier arrival and holding times.

Let A be the *link-route incidence matrix*,

$$A_{jr} = \begin{cases} 1, & j \in r, \\ 0, & j \notin r. \end{cases}$$

Remark 3.1 Often the set of links comprising a route will form a path between two nodes for some underlying graph, but we do not make this a requirement: for example, for a conference call the set of links may instead form a tree, or a complete subgraph, in an underlying graph. Later, from Section 3.3, we shall allow entries of A to be arbitrary non-negative integers, with the interpretation that a call requesting route r requires A_{jr} circuits from link j, and is lost if on any link $j = 1, \ldots, J$ there are fewer than A_{jr} circuits free. But for the moment assume A is a 0–1 matrix, and conveys the same information as the relation $j \in r$.

Let n_r be the number of calls in progress on route r. The number of circuits busy on link j is given by $\sum_r A_{jr} n_r$. Let $n = (n_r, r \in \mathcal{R})$, and let $C = (C_j, j \in \mathcal{J})$. Then n is a Markov process with state space

$$\{n \in \mathbb{Z}_+^R : An \leq C\}$$

(the inequality is to be read componentwise). This process is called a *loss network with fixed routing*.

Later, we will compute the exact equilibrium distribution for n. But first we describe a widely used approximation procedure.

Exercise

Exercise 3.1 Suppose that $\mu_k = 1, k = 1, \ldots, K$ in the model of Example 2.24. Show that the model is a loss network, with the resource set $\mathcal{J}$ taken as the set of edges in the interference graph.

3.2 Approximation procedure

The idea underlying the approximation is simple to explain. Let B_j be the probability of blocking on link j. Suppose that a Poisson stream of rate ν_r is thinned by a factor $1 - B_i$ at each link $i \in r \setminus j$ before being offered to link j. If these thinnings could be assumed independent both from link to link and over all routes passing through link j (they clearly are not), then the traffic offered to link j would be Poisson at rate

$$\sum_r A_{jr}\nu_r \prod_{i \in r \setminus \{j\}} (1 - B_i)$$

(the *reduced load*), and the blocking probability at link j would be given by Erlang's formula:

$$B_j = E\left(\sum_r A_{jr}\nu_r \prod_{i \in r - \{j\}} (1 - B_i), C_j\right), \qquad j = 1, \ldots, J. \tag{3.1}$$

Does a solution exist to these equations? First, let's recall (or note) the Brouwer fixed point theorem: a continuous map from a compact, convex set to itself has at least one fixed point. But the right-hand side of (3.1) defines a continuous map $F : [0, 1]^J \to [0, 1]^J$ via

$$(B_j,\ j = 1, \ldots, J) \mapsto \left(E\Big(\sum_r A_{jr}\nu_r \prod_{i \in r - \{j\}} (1 - B_i), C_j\Big),\ j = 1, \ldots, J\right),$$

and $[0, 1]^J$ is convex and compact. Hence there exists a fixed point. We shall see later that this fixed point is unique, and we call it, i.e. the solution to equations (3.1), the *Erlang fixed point*.

Example 3.2 Consider the following network with three cities linked via a transit node in the middle. We are given the arrival rates $\nu_{12}, \nu_{23}, \nu_{31}$. (We

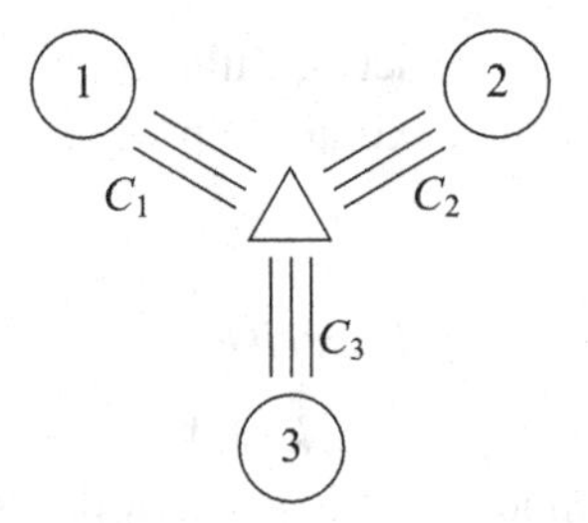

Figure 3.2 Telephone network for three cities.

write ν_{12} for the more cumbersome $\nu_{\{1,2\}}$). The approximation reads

$$\begin{aligned} B_1 &= E(\nu_{12}(1 - B_2) + \nu_{31}(1 - B_3), C_1), \\ B_2 &= E(\nu_{12}(1 - B_1) + \nu_{23}(1 - B_3), C_2), \\ B_3 &= E(\nu_{23}(1 - B_2) + \nu_{31}(1 - B_1), C_3), \end{aligned}$$

with corresponding loss probability along the route $\{1, 2\}$ given by

$$L_{12} = 1 - (1 - B_1)(1 - B_2),$$

and similarly for L_{23}, L_{31}.

Remark 3.3 How can we solve equations (3.1)? If we simply iterate the transformation F, we usually converge to the fixed point. This method, called *repeated substitution*, is often used in practice. If we use a sufficiently damped iteration, i.e. take a convex combination of the previous point and its image under F as the next point, we are guaranteed convergence. The complication of damping can be avoided by a variant of repeated substitution whose convergence you will establish later, in Exercise 3.7.

The approximation is often surprisingly accurate. Can we provide any insight into why it works well when it does? This will be a major aim in this chapter.

3.3 Truncating reversible processes

We now determine a precise expression for the equilibrium distribution of a loss network with fixed routing.

Consider a Markov process with transition rates $(q(j,k),\ j,k \in \mathcal{S})$. Say that it is *truncated* to a set $\mathcal{A} \subset \mathcal{S}$ if $q(j,k)$ is changed to 0 for $j \in \mathcal{A}$, $k \in \mathcal{S} \setminus \mathcal{A}$, and if the resulting process is irreducible within $\mathcal{A}$.

Lemma 3.4 *If a reversible Markov process with state space $\mathcal{S}$ and equilibrium distribution $(\pi(j),\ j \in \mathcal{S})$ is truncated to $\mathcal{A} \subset \mathcal{S}$, the resulting Markov process is reversible and has equilibrium distribution*

$$\pi(j)\left(\sum_{k\in\mathcal{A}}\pi(k)\right)^{-1}, \qquad j \in \mathcal{A}. \tag{3.2}$$

Proof By the reversibility of the original process

$$\pi(j)q(j,k) = \pi(k)q(k,j),$$

and so the probability distribution (3.2) satisfies detailed balance. □

Remark 3.5 If the original process is not reversible, then (3.2) is the equilibrium distribution of the truncated process if and only if

$$\pi(j)\sum_{k\in\mathcal{A}}q(j,k) = \sum_{k\in\mathcal{A}}\pi(k)q(k,j).$$

Earlier, equations of this form were termed "partial balance".

Consider a loss network with fixed routing for which $C_1 = \ldots = C_J = \infty$. If this is the case, arriving calls are never blocked; they simply arrive (at rate ν_r), remain for an exponentially distributed amount of time with unit mean, and leave. Henceforth allow the link-route incidence matrix A to have entries in $\mathbb{Z}_+$, not just 0 or 1.

This system is described by a linear migration process with transition rates

$$q(n, T_{\to r}n) = \nu_r, \qquad q(n, T_{r\to}n) = n_r,$$

and equilibrium distribution

$$\prod_{r\in\mathcal{R}} e^{-\nu_r}\frac{\nu_r^{n_r}}{n_r!}, \qquad n \in \mathbb{Z}^R.$$

(Since the capacities are infinite, the individual routes become independent.) If we now truncate n to $\mathcal{S}(C) = \{n : An \le C\}$, we obtain precisely the original loss network with fixed routing (and finite capacities). Therefore, its equilibrium distribution is

$$\pi(n) = G(C)\prod_r \frac{\nu_r^{n_r}}{n_r!}, \qquad n \in \mathcal{S}(C) = \{n : An \le C\}, \tag{3.3}$$

with

$$G(C) = \left(\sum_{n\in\mathcal{S}(C)}\prod_r \frac{\nu_r^{n_r}}{n_r!}\right)^{-1}.$$

Further, the equilibrium probability that a call on route r will be accepted is

$$1 - L_r = \sum_{n \in S(C - Ae_r)} \pi(n) = \frac{G(C)}{G(C - Ae_r)},$$

where $e_r \in S(C)$ is the unit vector that describes one call in progress on route r.

Remark 3.6 This is not a directly useful result for large (C big) or complex ($\mathcal{R}$ big) networks, because of the difficulty of computing the normalizing constant (in the case of an arbitrary matrix A, this is an NP-hard problem). However, we might hope for a limit result. We know that for large ν and C the Poisson distribution is going to be well approximated by a multivariate normal. Conditioning a multivariate normal on an inequality will have one of two effects: if the centre is on the feasible side of the inequality, the constraint has very little effect; if the centre is on the infeasible side of the inequality, we effectively restrict the distribution to the boundary of the constraint (because the tail of the normal distribution dies off very quickly), and the restriction of a multivariate normal to an affine subspace is again a multivariate normal distribution. In later sections we make this more precise.

We end this section with some further examples of truncated processes. These examples show that a variety of models may reduce to be equivalent to a loss network with fixed routing (and in particular the equilibrium distribution is given by independent Poisson random variables conditioned on a set of linear inequalities).

Example 3.7 (Call repacking) Consider the network in Figure 3.3 joining three nodes. Suppose that calls can be rerouted, even while in progress, if

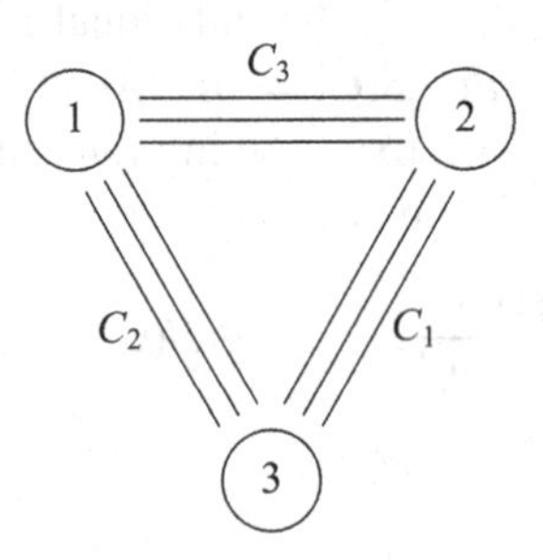

Figure 3.3 Loss network with call repacking.

this will allow another call to be accepted. (We also, as usual, assume that arrivals are Poisson and holding times are exponential.)

Let $n_{\alpha\beta}$ be the number of calls in progress between α and β. Then $n = (n_{12}, n_{23}, n_{31})$ is Markov, since all we need to know is the number of calls between each pair of cities. Further, we claim $n = (n_{12}, n_{23}, n_{31})$ is a linear migration process with equilibrium distribution

$$\prod_r e^{-\nu_r} \frac{\nu_r^{n_r}}{n_r!}, \quad r = \{1,2\},\ \{2,3\},\ \{3,1\}$$

truncated to the set

$$\mathcal{A} = \{n : n_{12} + n_{23} \le C_3 + C_1,\ n_{23} + n_{31} \le C_1 + C_2,\ n_{31} + n_{12} \le C_2 + C_3\}.$$

Indeed, it is clear that these restrictions are necessary: that is, if the state n is not in $\mathcal{A}$ then it cannot be reached. (Each of the three inequalities is a *cut constraint* separating one of the vertices from the rest of the network.) Next we show they are sufficient: that is, if $n \in \mathcal{A}$ then it is feasible to pack the calls so that all the calls in n are carried. First note that if all three inequalities of the form $n_{23} \le C_1$ hold (one for each direct route), the state n is clearly feasible. If not, suppose, without loss of generality, $n_{23} > C_1$. Then we can route C_1 of the calls directly and reroute the remaining $n_{23} - C_1$ calls via node 1. This will be possible provided $n_{23} - C_1 \le \min(C_3 - n_{12}, C_2 - n_{31})$, i.e. provided $n_{12} + n_{23} \le C_1 + C_3$ and $n_{23} + n_{31} \le C_1 + C_2$, which will necessarily hold since $n \in \mathcal{A}$. Thus in this case too the state n is feasible.

Thus, the equilibrium distribution is

$$\pi(n) = G(\mathcal{A}) \prod_r \frac{\nu_r^{n_r}}{n_r!}, \quad n \in \mathcal{A},\ r = \{1,2\},\ \{2,3\},\ \{3,1\}.$$

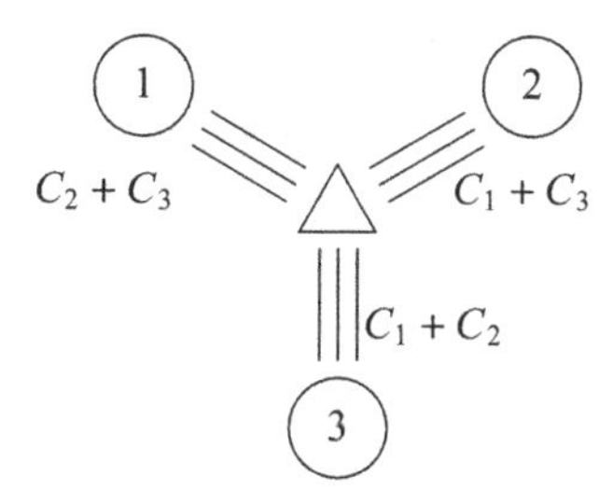

Figure 3.4 An equivalent network to the repacking model.

Note that the process n is equivalent to the loss network with fixed routing illustrated in Figure 3.4: the transition rates, as well as the state space and the equilibrium distribution, are identical.

Example 3.8 (A cellular network) Consider a cellular network, with seven cells arranged as in Figure 3.5. There are C distinct radio commu-

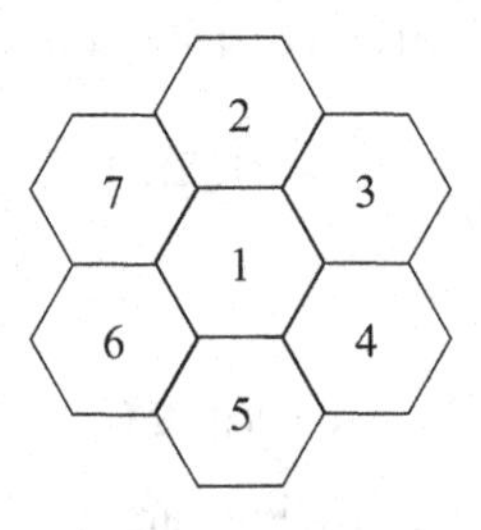

Figure 3.5 A cellular network: adjacent cells cannot use the same radio channel.

nication channels, but two adjacent cells cannot use the same channel. (A channel may be a frequency or a time slot, and interference prevents the same channel being used in adjacent cells. Two cells are adjacent in Figure 3.5 if they share an edge.) Calls may be reallocated between channels, even while they are in progress, if this will allow another call to be accepted. It is clear that $n_\alpha + n_\beta + n_\gamma \leq C$ for all cells α, β, γ that meet at a vertex in Figure 3.5. In Exercise 3.3 it is shown that these are the only constraints. Thus the network is equivalent to a loss network with fixed routing: it is as if there is a virtual link of capacity C associated with each vertex, and a call arriving at a cell requires one circuit from each of the virtual links associated with the vertices of that cell.

Exercises

Exercise 3.2 Let $m = C - An$, so that m_j is the number of free circuits on link j, and let $\pi'(m) = \sum_{n:An=C-m} \pi(n)$ be the equilibrium probability of the event that m_j circuits are free on link j for $j \in \mathcal{J}$. Establish the Kaufman–Dziong–Roberts recursion

$$(C_j - m_j)\pi'(m) = \sum_{r:Ae_r \leq C-m} A_{jr}\nu_r \pi'(m + Ae_r).$$

[*Hint:* Calculate $\mathbb{E}[n_r \mid m]$, using the detailed balance condition for $\pi(n)$. Note that the recursion can be used to solve for $\pi'(m)$ in terms of $\pi'(C)$, using the boundary condition that $\pi'(m) = 0$ if $m_j > C_j$ for any j.]

Exercise 3.3 Establish the result claimed in Example 3.8: that there is an allocation of radio channels to cells such that no channel is used in two adjacent cells if and only if $n_\alpha + n_\beta + n_\gamma \le C$ for all cells α, β, γ that meet at a vertex in Figure 3.5.
[*Hint:* Begin by allocating channels $1, \dots, n_1$ to cell 1, and channels $n_1 + 1, \dots, n_1 + n_2$ to cell 2, $n_1 + 1, \dots, n_1 + n_4$ to cell 4, and $n_1 + 1, \dots, n_1 + n_6$ to cell 6. Generalizations of this example are treated in Pallant and Taylor (1995) and Kind *et al.* (1998).]

3.4 Maximum probability

To obtain some insight into the probability distribution (3.3), we begin by looking for its mode, i.e. the location of the maximum of $\pi(n)$ over $n \in \mathbb{Z}_+^R$ and $An \le C$. Write

$$\log \pi(n) = \log\left(G(C) \prod_r \frac{\nu_r^{n_r}}{n_r!} \right) = \log G(C) + \sum_r (n_r \log \nu_r - \log(n_r!)).$$

By Stirling's approximation,

$$n! \sim \sqrt{2\pi} n^{n+\frac{1}{2}} e^{-n} \qquad \text{as } n \to \infty,$$

and so

$$\log(n!) = n \log n - n + O(\log n).$$

We will now replace the discrete variable n by a continuous variable x, and also ignore the $O(\log n)$ terms to obtain the following problem.

PRIMAL:

$$\begin{aligned} &\text{maximize} \sum_r (x_r \log \nu_r - x_r \log x_r + x_r) \\ &\text{subject to } Ax \le C \\ &\text{over} \qquad x \ge 0. \end{aligned} \tag{3.4}$$

The optimum in this problem is attained, since the feasible region is compact. The strong Lagrangian principle holds (Appendix C) since the objective function is concave, and the feasible region is defined by a set of linear inequalities. The Lagrangian is

$$L(x, z; y) = \sum_r (x_r \log \nu_r - x_r \log x_r + x_r) + \sum_j y_j \left(C_j - \sum_r A_{jr} x_r - z_j \right).$$

Here, $z \geq 0$ are the slack variables in the constraint $Ax \leq C$ and y are Lagrange multipliers. Rewrite as

$$L(x, z; y) = \sum_r x_r + \sum_r x_r \left(\log \nu_r - \log x_r - \sum_j y_j A_{jr} \right) + \sum_j y_j C_j - \sum_j y_j z_j,$$

which we attempt to maximize over $x, z \geq 0$. To obtain a finite maximum when we maximize the Lagrangian over $z \geq 0$, we require $y \geq 0$; and given this, we have at the optimum $y \cdot z = 0$. Further, differentiating with respect to x_r yields

$$\log \nu_r - \log x_r - \sum_j y_j A_{jr} = 0,$$

and hence at the maximum

$$x_r = \nu_r e^{-\sum_j y_j A_{jr}}.$$

Thus, if for given y we maximize the Lagrangian over $x, z \geq 0$ we obtain

$$\max_{x,z \geq 0} L(x, z; y) = \sum_r \nu_r e^{-\sum_j y_j A_{jr}} + \sum_j y_j C_j,$$

provided $y \geq 0$. This gives the following as the Lagrangian dual problem.

DUAL:

$$\text{minimize} \quad \sum_r \nu_r e^{-\sum_j y_j A_{jr}} + \sum_j y_j C_j \tag{3.5}$$

$$\text{over} \quad y \geq 0.$$

By the strong Lagrangian principle, there exists $\bar{y}$ such that the Lagrangian form $L(x, z; \bar{y})$ is maximized at $x = \bar{x}$, $z = \bar{z}$ that are feasible (i.e. $\bar{x} \geq 0$, $\bar{z} = C - A\bar{x} \geq 0$), and $\bar{x}$, $\bar{z}$ are then optimal for the PRIMAL problem. Necessarily, $\bar{y} \geq 0$ (dual feasibility) and $\bar{y} \cdot \bar{z} = 0$ (complementary slackness).

Let

$$e^{-y_j} = 1 - B_j.$$

We can rewrite the statement that there exist $\bar{y}$, $\bar{z}$ satisfying the conditions $\bar{y} \geq 0$ and $\bar{y} \cdot \bar{z} = 0$ in terms of the transformed variables as follows: there exist $(B_1, \ldots, B_J) \in [0, 1)^J$ that satisfy

Conditions on B:

$$\begin{cases} \sum_r A_{jr} \nu_r \prod_i (1 - B_i)^{A_{ir}} = C_j & \text{if } B_j > 0, \\ \qquad\qquad\qquad\qquad \leq C_j & \text{if } B_j = 0. \end{cases} \tag{3.6}$$

(To check the correspondence with the complementary slackness conditions, note that $B_j > 0$ if and only if $y_j > 0$. If you don't know the strong Lagrangian principle, don't worry: in a moment we'll establish directly the existence of a solution to the conditions on B.)

Remark 3.9 The conditions on B have an elegant fluid flow interpretation. With the substitution $e^{-y_j} = 1 - B_j$, the flow $\bar{x}_r = \nu_r \prod_i (1 - B_i)^{A_{ir}}$ looks like a thinning of the arrival stream ν_r by a factor $(1 - B_i)$ for each circuit requested from link i for each link i that the route r goes through. Then $\sum_r A_{jr} \nu_r \prod_i (1 - B_i)^{A_{ir}}$ is the aggregated flow on link j. The conditions on B tell us that the aggregated flow on link j does not exceed the link's capacity, and that blocking only occurs on links that are at full capacity.

Let's summarize what we have shown thus far, with a few additional points.

Theorem 3.10 *There exists a unique optimal solution $\bar{x} = (\bar{x}_r, r \in \mathcal{R})$ to the* Primal *problem* (3.4). *It can be expressed in the form*

$$\bar{x}_r = \nu_r \prod_j (1 - B_j)^{A_{jr}}, \quad r \in \mathcal{R},$$

where $B = (B_1, \ldots, B_J)$ is any solution to (3.6), *the conditions on B. There always exists a vector B satisfying these conditions; it is unique if A has rank J. There is a one-to-one correspondence between vectors satisfying the conditions on B and optima of the* Dual *problem* (3.5), *given by*

$$1 - B_j = e^{-\bar{y}_j}, \quad j = 1, \ldots, J.$$

Proof Strict concavity of the Primal objective function gives uniqueness of the optimum $\bar{x}$. The form of the optimum $\bar{x}$ was found above in terms of the Lagrange multipliers y, or equivalently B.

The explicit form of the Dual problem will allow us to establish directly the existence of B satisfying the conditions on B, without relying on the strong Lagrangian principle. Note that in the Dual problem we are minimizing a convex, differentiable function over the positive orthant, and the function grows to infinity as $y_j \to \infty$ for each $j = 1, \ldots, J$. Therefore, the function achieves its minimum at some point $\bar{y}$ in the positive orthant. If the minimum $\bar{y}$ has $\bar{y}_j > 0$, the partial derivative $\partial/\partial y_j|_{\bar{y}} = 0$. If $\bar{y}_j = 0$, the the partial derivative must be non-negative. (See Figure 3.6.)

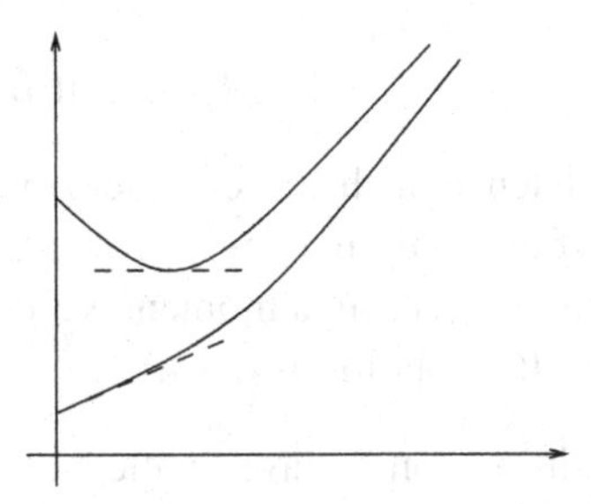

Figure 3.6 Minimum of a convex function over non-negative values occurs either where the derivative is zero, or at 0 with the derivative non-negative.

But this partial derivative of the DUAL objective function is simply

$$-\sum_r A_{jr}\nu_r e^{-\sum_i y_i A_{ir}} + C_j,$$

and so the minimum over $y \geq 0$ will be at a point where

$$\begin{cases} \sum_r A_{jr}\nu_r e^{-\sum_i y_i A_{ir}} & = C_j \qquad \text{if } y_j > 0, \\ & \leq C_j \qquad \text{if } y_j = 0. \end{cases}$$

Note that these are precisely the conditions on B, under the substitution $B_j = 1 - e^{-y_j}$; this gives the existence of a solution to the conditions on B, and establishes the one-to-one correspondence with the optima of the DUAL problem.

Finally, note that the DUAL objective function $\sum_r \nu_r e^{-\sum_j y_j A_{jr}}$ is strictly convex in the components of yA. If A has rank J, the mapping $y \mapsto yA$ is one-to-one, and therefore the objective function is also strictly convex in the components of y, and so there is a unique optimum. □

Example 3.11 (Rank deficiency) If A is rank deficient, the DUAL objective function may not be strictly convex in the components of y, and the optimizing y may not be unique. Consider

$$A = \begin{pmatrix} 1 & 0 & 0 & 1 \\ 1 & 1 & 0 & 0 \\ 0 & 1 & 1 & 0 \\ 0 & 0 & 1 & 1 \end{pmatrix}$$

corresponding to the system in Figure 3.7. The routes in $\mathcal{R}$ are $\{1,2\}, \{2,3\}$,

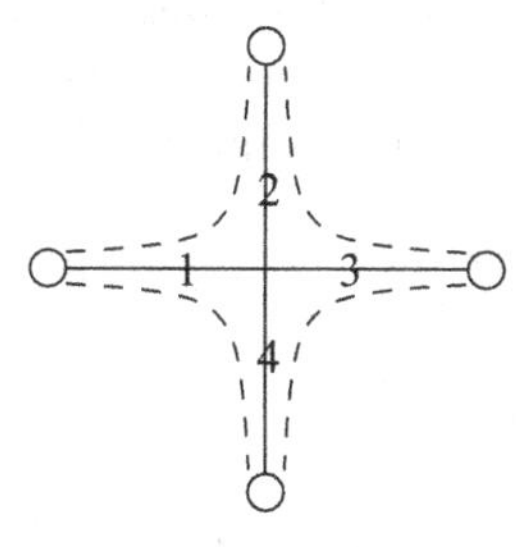

Figure 3.7 A network with a rank-deficient matrix.

$\{3,4\}, \{4,1\}$, and the matrix A has rank 3. In Exercise 3.4 it is shown that a solution to the conditions on B will, in general, be non-unique.

Remark 3.12 A will have rank J if there is some single-link traffic on each link (i.e. there exists a route $r = \{j\}$ for each $j \in \mathcal{R}$), since then the matrix A will contain the columns of the $J \times J$ identity matrix amongst its columns. This is a natural assumption in many cases.

Exercise

Exercise 3.4 Consider Example 3.11. Show that if $B_1, B_2, B_3, B_4 > 0$ solve the conditions on B, then so do

$$1 - d(1 - B_1), 1 - d^{-1}(1 - B_2), 1 - d(1 - B_3), 1 - d^{-1}(1 - B_4),$$

i.e. only $(1 - B_{odd})(1 - B_{even})$ is fixed. Deduce that for this example, where the matrix A is rank deficient, a solution to the conditions on B may be non-unique.

3.5 A central limit theorem

We noted earlier that we expect the truncated multivariate Poisson distribution (3.3) to approach a conditioned multivariate normal distribution as arrival rates and capacities become large. In this section we make this precise, and we shall see that the solutions to the optimization problems of Section 3.4 play a key role in the form of the limit.

We need first to define a limiting regime.

3.5.1 A limiting regime

Consider a sequence of networks as follows: replace $\nu = (\nu_r, r \in \mathcal{R})$ and $C = (C_j, j \in \mathcal{J})$ by $\nu(N) = (\nu_r(N), r \in \mathcal{R})$ and $C(N) = (C_j(N), j \in \mathcal{J})$ for $N = 1, 2, \ldots$. We will consider the scaling regime in which

$$\frac{1}{N}\nu_r(N) \to \nu_r, \qquad \text{for } r \in \mathcal{R},$$
$$\frac{1}{N}C_j(N) \to C_j, \qquad \text{for } j \in \mathcal{J},$$

as $N \to \infty$, where $\nu_r > 0$, $C_j > 0$ for all r and j. Write $B(N)$, $\bar{x}(N)$, etc. for quantities defined for the Nth network. Thus $B(N)$ solves the conditions on B (3.6) with ν, C replaced by $\nu(N), C(N)$.

From now on we will assume that A has full rank J, and moreover that the unique solution to the conditions on B (3.6) satisfies

$$\sum_r A_{jr}\nu_r \prod_i (1 - B_i)^{A_{ir}} < C_j \qquad \text{if } B_j = 0,$$

i.e. the inequality is strict. This will simplify the statements and proofs of results.

Lemma 3.13 *As $N \to \infty$, $B(N) \to B$ and $\frac{1}{N}\bar{x}(N) \to \bar{x}$.*

Proof The sequence $B(N)$, $N = 1, 2, \ldots$ takes values in the compact set $[0, 1]^J$. Select a convergent subsequence, say $B(N_k), k = 1, 2, \ldots$, and let $B' = \lim_{k\to\infty} B(N_k)$. Then for $N = N_k$ large enough,

$$\sum_r A_{jr}\nu_r(N) \prod_i (1 - B_i(N))^{A_{ir}} \begin{cases} = C_j(N), & \text{if } B'_j > 0, \\ \leq C_j(N), & \text{if } B'_j = 0. \end{cases}$$

(Note that this is true for all k with $B_j(N_k)$ on the right-hand side, but $B'_j > 0 \implies B_j(N_k) > 0$ for all k large enough and all $j = 1, \ldots, J$.)

Divide by N_k, and let $N_k \to \infty$, to obtain

$$\sum_r A_{jr}\nu_r \prod_i (1 - B'_i)^{A_{ir}} \begin{cases} = C_j, & \text{if } B'_j > 0, \\ \leq C_j, & \text{if } B'_j = 0. \end{cases}$$

Note that this shows $B'_j \neq 1$ for any j (it can't be equal to 1 in the second line, and it can't be equal to 1 in the first line because $C_j > 0$ for all j).

By the uniqueness of solutions to this set of relations, $B' = B$.

To finish the proof that $B(N) \to B$, we use a standard analysis argument. We have shown that any convergent sequence of $B(N)$ converges to B, and since we are in a compact set, any infinite subset of $\{B(N)\}$ has a convergent subsequence. Consider an open neighbourhood O of B; we show that all but finitely many terms of $B(N)$ lie in O. Indeed, the set $[0,1]^J \setminus O$ is still compact; if infinitely many terms $B(N_k)$ were to lie in it, they would have a convergent subsequence; but that convergent subsequence would (by above) converge to B, a contradiction.

Finally, since

$$\overline{x}_r(N) = \nu_r(N) \prod_i (1 - B_i(N))^{A_{ir}},$$

we conclude that $\overline{x}_r(N)/N \to \overline{x}_r$. □

We have shown that, in the limiting regime of high arrival rates and large capacities, we can approximate the most likely state of a large network by the most likely state of the limit. Our next task is to show that the distribution of $n(N)$, appropriately scaled, converges to a normal distribution conditioned on belonging to a subspace. The definitions below will be useful in defining the subspace.

Let

$$p_N(n(N)) = \prod_r \frac{\nu_r(N)^{n_r(N)}}{n_r(N)!}$$

(the unnormalized distribution of $n(N)$), and also let

$$m_j(N) = C_j(N) - \sum_r A_{jr} n_r(N),$$

the number of spare circuits on link j in the Nth network. Let

$$\mathcal{B} = \{j : B_j > 0\}, \quad A_{\mathcal{B}} = (A_{jr},\ j \in \mathcal{B}, r \in \mathcal{R}).$$

The matrix $A_{\mathcal{B}}$ contains those rows of A that correspond to links with positive entries of $\mathcal{B}$. Intuitively, those are the links where we expect to have positive blocking probability.

3.5.2 Limit theorems

Choose a sequence of states $n(N) \in \mathcal{S}(N)$, $N = 1, 2, \ldots$, and let

$$u_r(N) = N^{-1/2}(n_r(N) - \overline{x}_r(N)), \quad r \in \mathcal{R}.$$

Thus we are centring the vector $(u_r(N),\ r \in \mathcal{R})$ on the approximate mode, as found in Section 3.5.1, and using the appropriate scaling for convergence to a normal distribution.

Theorem 3.14 *The distribution of $u(N) = (u_r(N),\ r \in \mathcal{R})$ converges to the distribution of the vector $u = (u_r,\ r \in \mathcal{R})$ formed by conditioning independent normal random variables $u_r \sim N(0, \overline{x}_r)$, $r \in \mathcal{R}$, on $A_{\mathcal{B}} u = 0$. Moments converge also, and hence*

$$\frac{1}{N}\mathbb{E}[n_r(N)] \to \overline{x}_r,\ r \in \mathcal{R}.$$

Remark 3.15 The proof proceeds by directly estimating the unnormalized probability mass $p_N(n(N))$, and using Stirling's formula in the form

$$n! = (2\pi n)^{1/2} \exp(n \log n - n + \theta(n)),$$

where $1/(12n+1) < \theta(n) < 1/(12n)$. We shall not go through the more tedious parts of the proof (controlling error terms in the tail of the distribution), but we sketch the key ideas.

Sketch of proof Restrict attention initially to sequences $n(N) \in \mathcal{S}(N)$, $N = 1, 2, \ldots$, such that $\sum_r u_r(N)^2$ is bounded above: then $n_r(N) \sim \overline{x}_r(N) \sim \overline{x}_r N$ as $N \to \infty$. By Stirling's formula,

$$p_N(n(N)) = \prod_r (2\pi n_r(N))^{-1/2} \cdot \exp\Bigg(\underbrace{\sum_r (n_r(N)\log \nu_r(N) - n_r(N)\log n_r(N) + n_r(N))}_{X} + O(N^{-1})\Bigg).$$

We now expand X:

$$\begin{aligned} X &= -\sum_r n_r(N) \log \frac{n_r(N)}{e\nu_r(N)} \\ &= \underbrace{-\sum_r n_r(N) \log \frac{\overline{x}_r(N)}{\nu_r(N)}}_{\text{first term}} \underbrace{- \sum_r n_r(N) \log \frac{n_r(N)}{e\overline{x}_r(N)}}_{\text{second term}}. \end{aligned}$$

The first term, by the construction of $\overline{x}$ as a solution to an optimization problem, is

$$\begin{aligned} \sum_r n_r(N) \sum_j \overline{y}_j(N) A_{jr} &= \sum_j \overline{y}_j(N) \sum_r A_{jr} n_r(N) \\ &= \sum_j \overline{y}_j(N)(C_j(N) - m_j(N)). \end{aligned}$$

To deal with the second term, we note from a Taylor series expansion that

$$(1 + a)(1 - \log(1 + a)) = 1 - \frac{1}{2}a^2 + o(a^2) \qquad \text{as } a \to 0,$$

and therefore

$$\begin{aligned} -n_r(N) \log \frac{n_r(N)}{e\overline{x}_r(N)} &= (\overline{x}_r(N) + u_r(N)N^{1/2})\left(1 - \log\left(1 + \frac{u_r(N)N^{1/2}}{\overline{x}_r(N)}\right)\right) \\ &= \overline{x}_r(N)\left(1 - \frac{1}{2}\left(\frac{u_r(N)N^{1/2}}{\overline{x}_r(N)}\right)^2 + o(N^{-1})\right) \\ &= \overline{x}_r(N) - \frac{u_r(N)^2}{2\overline{x}_r} + o(1) \end{aligned}$$

(we absorb the error from replacing $\overline{x}_r(N)/N$ by $\overline{x}_r$ into the $o(1)$ term). Putting these together,

$$X = \sum_j \overline{y}_j(N)C_j(N) + \sum_r \overline{x}_r(N) - \sum_j \overline{y}_j(N)m_j(N) - \sum_r \frac{u_r(N)^2}{2\overline{x}_r} + o(1),$$

and thus

$$\begin{aligned} &\underbrace{p_N(n(N)) \exp\left(-\sum_r \overline{x}_r(N) - \sum_j \overline{y}_j(N)C_j(N)\right)}_{\text{term 1}} \\ &= \underbrace{\prod_j (1 - B_j(N))^{m_j(N)}}_{\text{term 2}} \underbrace{\prod_r (2\pi n_r(N))^{-1/2}}_{\text{term 3}} \underbrace{\exp\left(-\frac{u_r(N)^2}{2\overline{x}_r} + o(1)\right)}_{\text{term 4}}. \end{aligned}$$

Remark 3.16 We have established this form uniformly over all sequences $n(N)$ such that $\sum_r u_r(N)^2$ is bounded above and thus uniformly over $u(N)$ in any compact set. To be sure that this is enough to determine the limit distribution, we need to ensure tightness for the distributions of $u(N)$: we need to know that probability mass does not leak to infinity as $N \to \infty$, so that we can compute the limit distribution by normalizing the above form to be a probability distribution. We also need to ensure that the probability mass outside of a compact set decays quickly enough that moments can be calculated from the limit distribution. It is possible to ensure this by crudely bounding the error terms in Stirling's formula and the Taylor series expansion, but we omit this step.

Note that term 1 is a constant that does not depend on n (it does depend on N, of course): it will be absorbed in the normalizing constant. Similarly, since $n_r(N) \sim \overline{x}_r N$, term 3 can be absorbed in the normalizing constant, with an error that can be absorbed into the $o(1)$ term. Term 4 looks promising: we are trying to relate the probability distribution of $u(N)$ to the form $\exp(-u_r^2/2\overline{x}_r)$ corresponding to a $N(0, \overline{x}_r)$ random variable. But what about term 2?

We will now show that term 2 has the effect of conditioning the normal distribution on the set $A_{\mathcal{B}}u = 0$. For $j \in \mathcal{B}$, i.e. $B_j > 0$, we consider the quantity

$$m_j(N) = C_j(N) - \sum_r A_{jr} n_r(N) = C_j(N) - \sum_r A_{jr} \left(\overline{x}_r(N) + u_r(N) N^{1/2}\right)$$

$$= \underbrace{C_j(N) - \sum_r A_{jr}\overline{x}_r(N)}_{= 0 \text{ for } N \text{ large enough}} - \left(\sum_r A_{jr} u_r(N)\right) N^{1/2}.$$

Since $m_j(N) \geq 0$, we deduce that $\sum_r A_{jr} u_r(N) \leq 0$ for $j \in \mathcal{B}$ and N large enough. Further, term 2 decays geometrically with $m_j(N)$, for each $j \in \mathcal{B}$. Thus the relative mass assigned by $p_N(n(N))$ to values of $m_j(N) > m$ decays to zero as m increases, and so the relative mass assigned by $p_N(n(N))$ to values of $-\sum_r A_{jr} u_r(N)$ larger than any given $\epsilon > 0$ decays to zero as N increases. Therefore, the normalized limiting distribution is concentrated on the set $A_{\mathcal{B}}u = 0$, as required.

□

Corollary 3.17 *For $r \in \mathcal{R}$, and as $N \to \infty$,*

$$L_r(N) \to L_r \equiv 1 - \prod_i (1 - B_i)^{A_{ir}}.$$

Here, $L_r(N)$ is the probability that a call on route r is lost (in network N).

Proof By Little's law,

$$\underbrace{(1 - L_r(N))\nu_r(N)}_{\lambda} \cdot \underbrace{1}_{W} = \underbrace{\mathbb{E}[n_r(N)]}_{L}.$$

Therefore,

$$1 - L_r(N) = \frac{\mathbb{E}[n_r(N)]}{\nu_r(N)} = \frac{\mathbb{E}[n_r(N)]/N}{\nu_r(N)/N} \to \frac{\overline{x}_r}{\nu_r} = \prod_i (1 - B_i)^{A_{ir}}.$$

□

Remark 3.18 Observe that if there is a link $r = \{j\}$, then the Corollary shows that $L_r(N) \to B_j$.

Example 3.19 Consider the system in Figure 3.8, with link-route incidence matrix

$$A = \begin{pmatrix} 1 & 0 & 1 \\ 0 & 1 & 1 \end{pmatrix}.$$

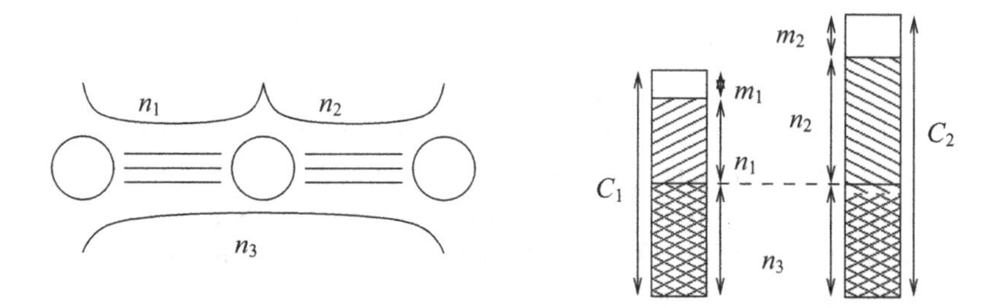

Figure 3.8 Shared communication link and breakdown of the number of circuits in use/free on each of the links.

Suppose $\mathcal{B} = \{1,2\}$, so $A_{\mathcal{B}} = A$. Then prior to conditioning on $A_{\mathcal{B}}u = 0$ we have $u_r = (n_r - \bar{x}_r)/\sqrt{N} \to N(0, \bar{x}_r)$, three independent normals. However, as $|\mathcal{B}| = 2$, the condition $A_{\mathcal{B}}u = 0$ reduces the support by two dimensions, constraining u to a one-dimensional subspace, where $n_1 + n_3 \approx C_1$ and $n_2 + n_3 \approx C_2$.

The free circuit processes m_1 and m_2 control the system: link j blocks calls if and only if $m_j = 0$. Thus m_1 and m_2 are often not zero. But, in the limiting regime, m_1 and m_2 are only $O(1)$, whereas n_1, n_2, n_3, C_1 and C_2 all scale as $O(N)$.

3.6 Erlang fixed point

Consider the equations

$$E_j = E\left((1 - E_j)^{-1} \sum_r A_{jr} \nu_r \prod_i (1 - E_i)^{A_{ir}}, C_j\right), \qquad j = 1, \ldots, J, \quad (3.7)$$

the generalization of the Erlang fixed point equations to matrices A that may not be 0–1. Our goal now is to show that there exists a unique solution to these equations, and that, in the limiting regime, the solution converges to the correct limit B, namely that arising from the maximum probability optimization problem.

Theorem 3.20 *There exists a unique solution $(E_1, \ldots, E_J) \in [0,1]^J$ satisfying the Erlang fixed point equations* (3.7).

Proof We prove this theorem by showing that we can rewrite the Erlang fixed point equations as the stationary conditions for an optimization problem (with a unique optimum).

Define a function $U(y, C) : \mathbb{R}_+ \times \mathbb{Z}_+ \to \mathbb{R}_+$ by the implicit relation

$$U(-\log(1 - E(\nu, C)), C) = \nu(1 - E(\nu, C)).$$

The interpretation is that $U(y, C)$ is the *utilization* or mean number of circuits in use in the Erlang model, when the blocking probability is $E = 1 - e^{-y}$. (See Exercise 3.5.) Observe that as ν increases continuously from 0 to ∞ the first argument of U increases continuously from 0 to ∞, and so this implicit relation defines a function $U(y, C) : \mathbb{R}_+ \times \mathbb{Z}_+ \to \mathbb{R}_+$. As both the utilization $\nu(1 - E(\nu, C))$ and the blocking probability $E(\nu, C)$ are strictly increasing functions of ν, the function $U(y, C)$ is a strictly increasing function of y. Therefore, the function

$$\int_0^y U(z, C)\, dz$$

is a strictly convex function of y.

Consider now the optimization problem

REVISED DUAL:

$$\text{minimize} \quad \sum_r \nu_r e^{-\sum_j y_j A_{jr}} + \sum_j \int_0^{y_j} U(z, C_j)\, dz \tag{3.8}$$

$$\text{over} \quad y \geq 0.$$

Note that this problem looks a lot like (3.5), except that we have replaced the linear term in the earlier objective function by a strictly convex term.

By the strict convexity of its objective function (3.8), the REVISED DUAL has a unique minimum. Differentiating, the stationary conditions yield that, at the unique minimum,

$$\sum_r A_{jr} \nu_r e^{-\sum_i y_i A_{ir}} = U(y_j, C_j), \qquad j = 1, \ldots, J. \tag{3.9}$$

Now suppose E solves the Erlang fixed point equations (3.7), and define y_j by $E_j = 1 - e^{-y_j}$ (i.e. $y_j = -\log(1 - E_j)$). We can rewrite (3.7) in terms of y as

$$E\left(e^{y_j} \sum_r A_{jr} \nu_r e^{-\sum_i y_i A_{ir}}, C_j\right) = 1 - e^{-y_j}, \qquad j = 1, \ldots, J,$$

or, moving things from side to side and multiplying by an arbitrary real parameter λ,

$$\lambda e^{-y_j} = \lambda \left(1 - E\left(e^{y_j} \sum_r A_{jr} \nu_r e^{-\sum_i y_i A_{ir}}, C_j \right) \right). \tag{3.10}$$

But if we make the choice

$$\lambda = e^{y_j} \sum_r A_{jr} \nu_r e^{-\sum_i y_i A_{ir}}$$

and use the definition of U, equations (3.10) will become precisely the statement of the stationary conditions (3.9). Since the solution to the stationary conditions is unique, we deduce that there exists a unique solution to the Erlang fixed point equations. □

We now show that the objective function of the REVISED DUAL asymptotically approaches that of the DUAL problem: first we show that $U(z, C_j)$ is close to C_j (except, of course, for $z = 0$).

Lemma 3.21

$$U(y, C) = C - (e^y - 1)^{-1} + o(1)$$

as $C \to \infty$, *uniformly over* $y \in [a, b] \subset (0, \infty)$ *(i.e.* y *on compact sets, bounded away from 0).*

Proof Consider an isolated Erlang link of capacity C offered Poisson traffic at rate ν. Then

$$\pi(j) = \frac{\nu^j}{j!} \left(\sum_{k=0}^{C} \frac{\nu^k}{k!} \right)^{-1}.$$

Let $\nu, C \to \infty$ with $C/\nu \to 1 - B$ for some $B > 0$. (That is, the capacity is smaller than the arrival rate by a constant factor, and the ratio gives B.) Then

$$\pi(C) = \frac{1}{1 + \frac{C}{\nu} + \frac{C(C-1)}{\nu^2} + \ldots} \to B,$$

since the denominator converges to $1 + (1 - B) + (1 - B)^2 + \ldots$.

Now suppose a more precise relationship between ν, C: suppose that, as $C \to \infty$, ν takes values so that $\pi(C) = B$ for fixed $B \in (0, 1)$. Then the expectation (with respect to π) of the number of free circuits is

$$\sum_{m=0}^{C} m\pi(C-m) = \pi(C)\left(0 + 1\cdot\frac{C}{\nu} + 2\cdot\frac{C(C-1)}{\nu^2} + \dots\right)$$

$$\le \pi(C)\sum_{m=0}^{\infty} m\frac{C^m}{\nu^m} = \pi(C)\frac{C/\nu}{(1-C/\nu)^2} \to \frac{B(1-B)}{B^2} = \frac{1-B}{B}.$$

Indeed, we have bounded convergence: the elements of the series converge term-by-term to the geometric bound, which is finite. This implies that in the limit we have equality in the place of $\le$; i.e.

$$\mathbb{E}_\pi[\text{number of free circuits}] \to \frac{1-B}{B} = \frac{e^{-y}}{1-e^{-y}} = (e^y-1)^{-1},$$

where, as usual, $B = 1 - e^{-y}$. Since the utilization $U(y, C)$ is the expected number of busy circuits when the blocking probability is $B = 1 - e^{-y}$, we have established the pointwise limit stated in the result. To deduce uniform convergence, observe that for B in a compact set contained within $(0, 1)$ the error in the bounded convergence step can be controlled uniformly over the compact set. □

Let $E_j(N)$ be the Erlang fixed point, i.e. the solution to (3.7), for network N in the sequence considered in Section 3.4. As in that section, assume A is of full rank, so that the solution to (3.6), the conditions on B, is unique.

Corollary 3.22 *As $N \to \infty$, $E_j(N) \to B_j$.*

Proof The Erlang fixed point is the unique minimum of the Revised dual objective function, and from Lemma 3.21 we can write this objective function as

$$\sum_r \nu_r(N)e^{-\sum_j y_j A_{jr}} + \sum_j \int_0^{y_j} U(z, C_j(N))dz$$

$$= N\left(\sum_r \nu_r e^{-\sum_j y_j A_{jr}} + \sum_j y_j C_j + o(1)\right),$$

where the convergence is uniform on compact subsets of $(0, \infty)^J$. That is, the Revised dual objective function, scaled by N, converges uniformly to the Dual objective function of (3.5). Since the Dual objective function is strictly convex and continuous, the minimum of the Revised dual objective function converges to it as $N \to \infty$, as required.

□

Remark 3.23 The corollary to Theorem 3.14 showed that the limiting loss probability L_r is as if links block independently, with link j rejecting a request for a circuit with probability B_j, where $(B_1, \ldots, B_J)$ is the unique solution to the conditions on B. Corollary 3.22 shows, reassuringly, that the Erlang fixed point converges to the same vector $(B_1, \ldots, B_J)$.

Exercises

Exercise 3.5 Let y and C be given. Consider a single Erlang link of capacity C, whose arrival rate $\nu = \nu(y, C)$ is such that the probability of a lost call is equal to $1 - e^{-y}$. Use Erlang's formula to relate y and ν. Then use Exercise 1.7 (or Exercise 2.15) to find the mean number of circuits in use, $U(y, C)$, and verify that it has the form given in the proof of Theorem 3.20.

Exercise 3.6 Consider the isolated Erlang link of Lemma 3.21 under the limiting regime considered there. Show that the distribution of the number of free circuits on the link converges to a geometric distribution, whose probability of being equal to 0 is the blocking probability B.

Exercise 3.7 In Section 3.2 repeated substitution was noted as a method for finding the Erlang fixed point. A variant of repeated substitution is to start from a vector B, perhaps $(0, 0, \ldots, 0)$, and use the right-hand side of (3.1) to update the components of B cyclically, one component at a time. Show that, when A is a 0–1 matrix, this corresponds to finding the minimum in each coordinate direction of the function (3.8) cyclically, one coordinate at a time.
[*Hint:* Show that solving the jth equation (3.9) for y_j, for given values of y_i, $i \neq j$, corresponds to finding the minimum of the function (3.8) in the jth coordinate direction.]

Deduce from the strict convexity of the continuously differentiable function (3.8) that this method converges.

Describe a generalization of this method, with proof of convergence, for the case when A is not a 0–1 matrix.

3.7 Diverse routing

The limit theorems of earlier sections concern networks with increasing link capacities and loads, but fixed network topology. Next we briefly consider a different form of limiting regime, where link capacities and loads are fixed or bounded and where the numbers of links and routes in the network increase to infinity.

We begin with a simple example. Consider the star network with n links, as in Figure 3.9. An arriving call requires a single circuit from each of two

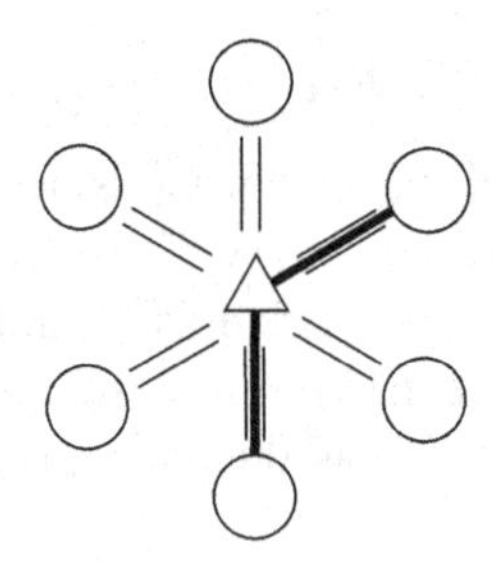

Figure 3.9 Star network with six links, and one of the routes.

randomly chosen links. Suppose we leave constant the capacities of links, but let the number of links increase to infinity, and suppose the total arrival rate of calls in the entire network increases at the same rate. Then it is possible to show that the Erlang fixed point emerges under the limit as $n \to \infty$.

Note that the probability that two calls sharing a link actually share their second link also tends to 0 as $n \to \infty$, as this will be true for a network with all capacities infinite, and at any time t the calls present in a finite capacity network are a subset of those present in the infinite capacity network (under the natural coupling).

The Erlang fixed point also emerges for various other network topologies, provided the routing is sufficiently diverse, in the sense that the probability approaches zero that two calls through a given link share a link elsewhere in the network. This is unsurprising (although non-trivial to prove), as we would expect the assumption of diverse routing to lead quite naturally to links becoming less dependent, and the Erlang fixed point is based on a link independence approximation.

The Erlang fixed point is appealing as an approximation procedure since it works well in a range of limiting regimes, and these regimes cover networks with large capacities and/or diverse routing, for which exact answers are hard to compute.

3.7.1 Non-uniqueness

We have seen that under fixed routing the Erlang fixed point is unique. Next we consider an example of alternative routing, interesting in its own right, where the fixed point emerges from a diverse routing limit, but is not unique.

Consider a network that is a complete graph on n nodes and symmetric: the arrival rate between each pair of nodes is ν, and the number of circuits on each of the links is C. Suppose that an arriving call is routed directly

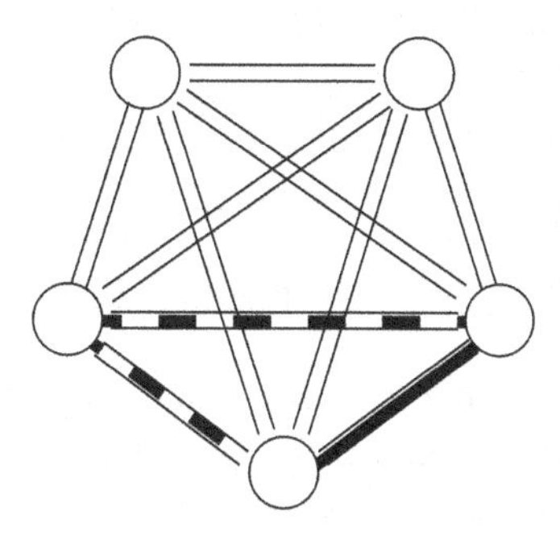

Figure 3.10 Complete graph topology, with five nodes. A route (solid) and an alternative route (dashed) are highlighted.

if possible, and otherwise a randomly chosen two-link alternative route is tried; if that route happens to be blocked at either link, the call is lost.

We will develop an Erlang fixed point equation for this model, based on the same underlying approximation that links block independently. In that case, if B is the blocking probability on a link, taken to be the same for each link, then

$$\mathbb{P}(\text{incoming call is accepted}) = \underbrace{(1-B)}_{\text{can route directly}} + \underbrace{B(1-B)^2}_{\substack{\text{can't route directly,} \\ \text{but can via two-link detour}}},$$

and the expected number of circuits per link that are busy is $\nu(1-B) + 2\nu B(1-B)^2$. Thus, we look for a solution to

$$B = E(\nu(1 + 2B(1-B)), C) \tag{3.11}$$

(since if an arrival rate $\nu(1 + 2B(1-B))$ is thinned by a factor $1 - B$ we get the desired expected number of busy circuits).

Remark 3.24 An alternative way to derive this arrival rate is simply to count the calls that come for link ij: we have an arrival rate of ν for the calls $i \leftrightarrow j$, plus for each other node k we have a traffic of $\nu B(1-B)/(n-2)$ for calls $i \leftrightarrow k$ that are rerouted via j (and the same for calls $j \leftrightarrow k$ that are rerouted via i). Adding these $2(n-2)$ terms gives the total arrival rate $\nu(1 + 2B(1 - B))$.

The solutions to the fixed point equation (3.11) are illustrated in Figure 3.11. The curve corresponding to $C = \infty$ arises as follows. Suppose that $\nu, C \to \infty$ while keeping their ratio fixed. Then

$$\lim_{N\to\infty} E(\nu N, CN) = (1 - C/\nu)^+,$$

where $x^+ = \max(x, 0)$ is the positive part of x. The fixed point equation (3.11) therefore simplifies to

$$B = [1 - C/\nu(1 + 2B(1 - B))]^+ ,$$

and the locus of points satisfying this equation is plotted as the dashed curve in Figure 3.11. Observe that for some values of ν and C there are multiple solutions (for example, when $C = \infty$ we have a cubic equation for B, which has multiple roots).

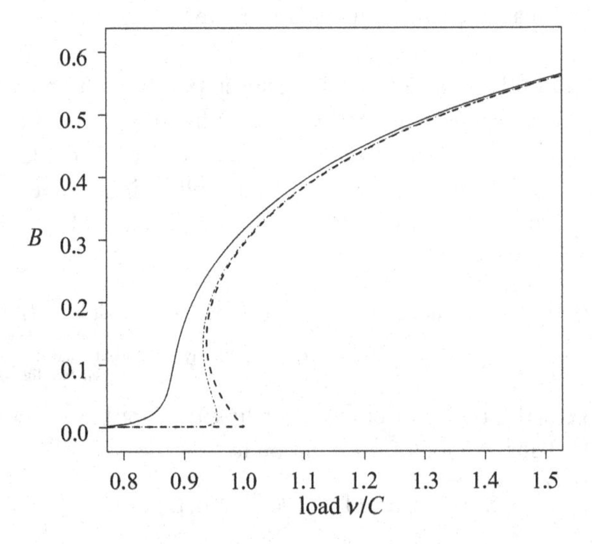

Figure 3.11 Blocking probabilities from the fixed point equation for $C = 100$ (solid), $C = 1000$ (dash-dotted) and the limiting $C = \infty$ (dashed).

Simulations of the system with n large show hysteresis, as in Figure 3.12. When the arrival rate is slowly increased from 0, the blocking probability

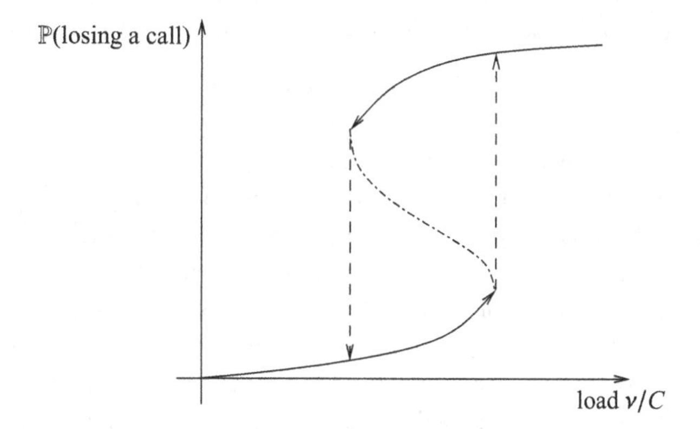

Figure 3.12 Sketch of loss probability, as we slowly increase or slowly decrease the arrival rate. The dashed curve sketches the solution of the Erlang fixed point equation on the same axes.

follows the lower solution for B until that has an infinite derivative; it then jumps to the higher value. If the arrival rate is then slowly decreased, the blocking probability follows the higher solution for B until that has an infinite derivative, then jumps to the lower value.

An intuitive explanation is as follows. The lower solution corresponds to a mode in which blocking is low, calls are mainly routed directly, and relatively few calls are carried on two-link paths. The upper solution corresponds to a mode in which blocking is high and many calls are carried over two-link paths. Such calls use two circuits, and this additional demand on network resources may cause a number of subsequent calls also to attempt two-link paths. Thus a form of positive feedback may keep the system in the high blocking mode.

And yet the system is an irreducible finite-state Markov chain, so it must have a *unique* stationary distribution, and hence a *unique* probability of a link being full, at any given arrival rate.

How can both these insights be true? We can indeed reconcile them, by observing that they concern two different scaling regimes:

(1) if we leave the arrival rate constant and simulate the network for long enough, the proportion of time a link is full or the proportion of lost calls will indeed converge to the stationary probability coming from the unique stationary distribution;

(2) if, however, we fix a time period $[0, T]$, fix the per-link arrival rate of calls, and let the number of links tend to infinity, the time to jump from one branch of the graph to the other will tend to infinity, i.e. the system will freeze in one of the two modes (low utilization or high utilization).

Remark 3.25 In case (1), the unique stationary distribution for the chain may be bimodal, with a separation between the two modes that becomes more pronounced as n increases. If you are familiar with the Ising model of a magnet, a similar phenomenon occurs there. The Ising model considers particles located at lattice points with states ± 1, which are allowed to switch states, and the particle is more likely to switch to a state where it agrees with the majority of its neighbours. The unique stationary distribution of any finite-sized system is symmetric with mean zero; on the other hand, if we look over a finite time horizon at ever-larger squares, for some parameter values the system will freeze into a mode where most particles are $+1$ or a mode where most particles are -1.

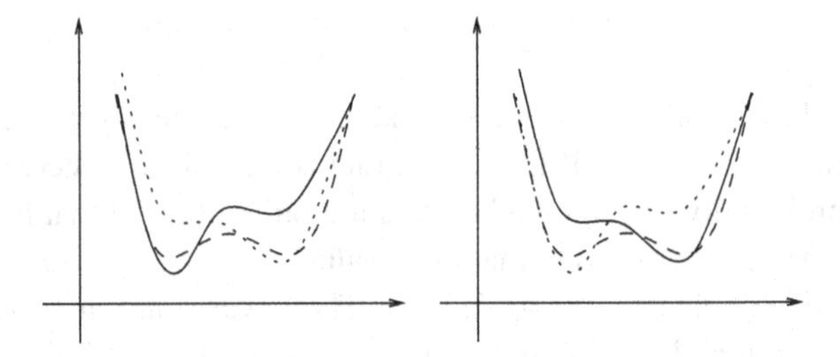

Figure 3.13 As we transition from the solid to the dashed and then the dotted curve, the location of the minimum changes abruptly. The value of the minimum will change continuously.

Remark 3.26 Some further insight into the instability apparent in this example can be obtained as follows. The fixed point equations (3.11) locate the stationary points of

$$\nu e^{-y} + \underbrace{\nu e^{-2y}(1 - 2/3e^{-y})}_{\text{non-convex!}} + \int_0^y U(z, C)\, dz, \tag{3.12}$$

where U is the utilization we defined in the proof of Theorem 3.20. (You will show this in Exercise 3.8.) Since this is a non-convex function, changing the parameters slightly can change the location of the stationary points quite a lot. For an example of this phenomenon, see Figure 3.13, which plots a family of non-convex curves (not this family, however).

The diverse routing regime can be used to provide a deeper understanding of this example. Let $Q_{nj}(t)$ be the number of links with j busy circuits at time t. Let $X_{nj}(t) = n^{-1}Q_{nj}(t)$, and let $X_n(t) = (X_{nj}, j = 0, 1, \ldots, C)$. Note that $(X_n(t), t \in [0, T])$ is a random function. But it converges to a deterministic function as $n \to \infty$. Indeed, it can be established (Crametz and Hunt (1991)) that if $X_n(0)$ converges in distribution to $X(0)$ then $(X_n(t), t \in [0, T])$ converges in distribution to $(X(t), t \in [0, T])$, where $X(\cdot) = (x_0(\cdot), x_1(\cdot), \ldots, x_C(\cdot))$ is the unique solution to the equations

$$\frac{d}{dt}\left(\sum_{i=0}^{j} x_i(t)\right) = (j+1)x_{j+1}(t) - (\nu + \sigma(t))x_j(t), \qquad j = 0, 1, \ldots, C-1, \tag{3.13}$$

where

$$\sigma(t) = 2\nu x_C(t)(1 - x_C(t)).$$

An intuition into these equations is as follows: the sum on the left is the proportion of links with j or fewer busy circuits; this will increase when a circuit in a link with $j+1$ busy circuits becomes free, and decrease when a call arrives at a link with j busy circuits.

Exercise 3.9 shows that if a probability distribution $x = (x_0, x_1, \ldots, x_C)$ is a fixed point of the above system of differential equations, then $x_C = B$, where B is a solution to the fixed point equation (3.11), and hence a stationary point of the function (3.12). The upper and lower solutions for B in Figure 3.11 correspond to minima in Figure 3.13 and stable fixed points of the differential equations, and the middle solution corresponds to the intervening maximum in Figure 3.13 and to an unstable fixed point of the differential equation.

3.7.2 Sticky random routing

Mismatches between traffic and capacity are common in communication networks, often caused by forecast errors or link failures. Can some form of alternative routing allow underutilized links to help with traffic overflowing from busy links? Section 3.7.1 has shown what can go wrong. And if we allow calls to try more and more alternative routes, the problem becomes *worse*, as a higher and higher proportion of calls are carried over two circuits. How can we control this, given that we *do* want to allow rerouting of calls? We describe a simple and effective method that was first implemented in the BT network and elsewhere in the 1990s. There are two key ideas.

First: *trunk reservation*. A link will accept an alternatively routed call only if there are $\geq s$ circuits on the link, where $s \ll C$ is a small constant (for example, $s = 5$ for $C = 100, 1000$). This is an idea with a long history, and achieves a form of priority for directly routed calls over alternatively routed calls when the link is busy.

Second: the following *sticky random scheme*. A call $i \to j$ is routed directly if there is a free circuit on the link between them. Otherwise, we try to reroute the call via a tandem node $k(i, j)$ (which is stored at i). If trunk reservation allows the rerouting, the call is accepted. If not, the call is lost, and the tandem node $k(i, j)$ is reset randomly. Note especially that the tandem node is not reselected if the call is successfully routed on either the direct link or the two-link alternative route. The intuition is that the system will look for links with spare capacity in the network, and that trunk reservation will discourage two-link routing except where there is spare capacity.

Dynamic routing strategies such as this are effective in allowing a network to respond robustly to failures and overloads. Good strategies effectively pool the resources of the network, so that spare capacity in part of the network can be available to deal with excess traffic elsewhere.

3.7.3 Resource pooling

Next we look at a very simple model to illustrate an interesting consequence of *resource pooling*. Consider a number of independent and parallel Erlang links, each with the same load and capacity. What could be gained by pooling them, so that a call for any of the links could use a circuit from any of them? Well, to calculate the benefit we simply need to recalculate Erlang's formula $E(\nu, C)$ with both its parameters increased by the same factor, the number of links pooled. The effect on the loss probability is shown in Figure 3.14. Intuitively, the aggregating of load and of capacity lessens the effect of randomness, and this reduces the loss probability, a phenomenon known in the telecommunications industry as *trunking efficiency*. (If the loads differed from link to link, the reduction in loss probability from pooling would be even greater.)

Thus, as the amount of resource pooling increases, that is as C increases, the blocking probability decreases for a given load ν/C. Indeed, in the limit as $C \to \infty$, the blocking probability approaches 0 for any load less than 1. This admirable state of affairs has one unfortunate consequence. Imagine the load on the network is gradually increasing, perhaps over months, and that the blocking probability is used as an indication of the health of the

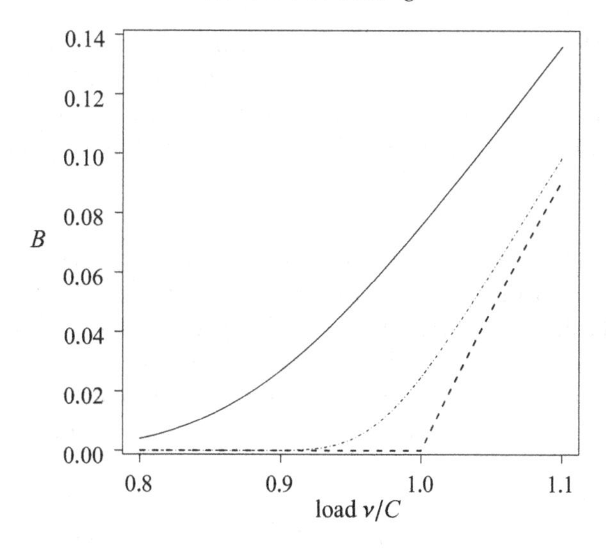

Figure 3.14 Erlang blocking probability for $C = 100$ (solid), $C = 1000$ (dot-dashed) and the limit as $C \to \infty$ (dashed).

network; in particular, suppose an increase in blocking is used to indicate a need for capacity expansion, which may take time to implement. When C is small the gradual increase in blocking as load increases gives plenty of time for capacity expansion. But when C is large there is a problem. In the limiting case of $C = \infty$, nothing is noticed until the load passes through 1: at this point the blocking probability curve has a discontinous derivative, and blocking increases rapidly. The moral of this discussion is that, in networks with very efficient resource pooling, the blocking probability alone is not a good measure of how close the system is to capacity, and additional information may be needed. We return to this point at the end of Section 4.3.

Exercises

Exercise 3.8 Show that the fixed point equations (3.11) locate the stationary points of the function (3.12).
[*Hint:* Use the definition of the function $U(\cdot)$ from the proof of Theorem 3.20.]

Exercise 3.9 Show that if a probability distribution $x = (x_0, x_1, \ldots, x_C)$ is a fixed point of the system of differential equations (3.13), then it is also the stationary distribution for an Erlang link of capacity C with a certain arrival rate. Deduce that necessarily $x_C = B$, where B is a solution to the fixed point equation (3.11).

Exercise 3.10 Suppose that when a call $i \to j$ is lost in the sticky random scheme the tandem node $k(i, j)$ is reset either at random from the set of nodes other than i and j, or by cycling through a random permutation of this set of nodes. Let $p_k(i, j)$ denote the long-run proportion of calls $i \to j$ that are offered to tandem node k, and let $q_k(i, j)$ be the long-run proportion of those calls $i \to j$ and offered to tandem node k that are blocked. Show that

$$p_a(i, j)q_a(i, j) = p_b(i, j)q_b(i, j), \qquad a, b \neq i, j.$$

Deduce that, if the blocking from $i \to j$ is high on the path via the tandem node k, the proportion of overflow routed via node k will be low.

Exercise 3.11 Consider pooling C independent $M/M/1$ queues, each with the same parameters, into a single $M/M/1$ queue with service rate C. Show that the mean sojourn time in the system of a customer is divided by C. Sketch a diagram parallel to Figure 3.14, plotting mean sojourn time against load. Note that, in the limiting case of $C = \infty$, the mean sojourn time is discontinuous at a load of 1.

Exercise 3.12 Trunk reservation has several other uses. Consider a single link of capacity C at which calls of type k, requiring A_k circuits, arrive as independent Poisson streams of rate ν_k. Let $n = (n_k)_k$ describe the number of calls of each type in progress. Suppose calls of type k are accepted if and only if the vector n prior to admission satisfies

$$\sum_r A_r n_r \leq C - s_k,$$

where the trunk reservation parameters s_k satisfy $s_k \geq A_k$. Note that varying s_k varies the relative loss probabilities for different types of call. Show that if, for each k, $s_k = \max_r A_r$, the loss probability does not depend upon the call type.

3.8 Further reading

Extensive reviews of work on loss networks are given by Kelly (1991) and Ross (1995); Zachary and Ziedins (2011) review recent work on dynamical behaviour. Sticky random schemes are surveyed by Gibbens *et al.* (1995).

Part II

4

Decentralized optimization

A major practical and theoretical issue in the design of communication networks concerns the extent to which control can be decentralized. Over a period of time the form of the network or the demands placed on it may change, and routings may need to respond accordingly. It is rarely the case, however, that there should be a central decision-making processor, deciding upon these responses. Such a centralized processor, even if it were itself completely reliable and could cope with the complexity of the computational task involved, would have its lines of communication through the network vulnerable to delays and failures. Rather, control should be decentralized and of a simple form: the challenge is to understand how such decentralized control can be organized so that the network as a whole reacts sensibly to fluctuating demands and failures.

The behaviour of large-scale systems has been of great interest to mathematicians for over a century, with many examples coming from physics. For example, the behaviour of a gas can be described at the microscopic level in terms of the position and velocity of each molecule. At this level of detail a molecule's velocity appears as a random process, with a stationary distribution as found by Maxwell. Consistent with this detailed microscopic description of the system is macroscopic behaviour, best described by quantities such as temperature and pressure. Similarly, the behaviour of electrons in an electrical network can be described in terms of random walks, and yet this simple description at the microscopic level leads to rather sophisticated behaviour at the macroscopic level: the pattern of potentials in a network of resistors is just such that it minimizes heat dissipation for a given level of current flow. The local, random behaviour of the electrons causes the network as a whole to solve a rather complex optimization problem.

Of course, simple local rules may lead to poor system behaviour if the rules are the wrong ones. Road traffic networks provide a chastening example of this. Braess's paradox describes how, if a new road is added to a congested network, the average speed of traffic may fall rather than rise, and indeed everyone's journey time may lengthen. The attempts of individuals to do the best for themselves lead to everyone suffering. It is possible to alter the local rules, by the imposition of appropriate tolls, so that the network behaves more sensibly, and indeed road traffic networks provided an early example of the economic principle that externalities need to be appropriately penalized for decentralized choices to lead to good system behaviour.

In this chapter, we discuss these examples of decentralized optimization. In Section 4.1 we discuss a simple model of the motion of electrons in a network of resistors, and in Section 4.2 we describe a model of road traffic. We shall see that our earlier treatment of loss networks in Chapter 3 can be placed in this more general context, and we shall work through some examples which present ideas we shall see again in later chapters on congestion control.

4.1 An electrical network

We begin with a description of a symmetric random walk on a graph, which we later relate to a model of an electrical network.

4.1.1 A game

Consider the following game. On a graph with vertex set G, you perform a symmetric random walk with certain transition rates until you hit a subset $S \subset V$. (A random walk on a graph with transition rates $\gamma = (\gamma_{jk})$ is a Markov process with state space G and transition rates $q_{jk} = \gamma_{jk}$ if (j, k) is an edge, and $q_{jk} = \gamma_{jk} = 0$ otherwise. A symmetric random walk has $\gamma_{jk} = \gamma_{kj}$.) The game ends when you reach a vertex in S, and you receive a reward that depends on the particular vertex you hit: if the vertex is i, the reward is v_i. How much should you pay to play this game? That is, what is your expected reward?

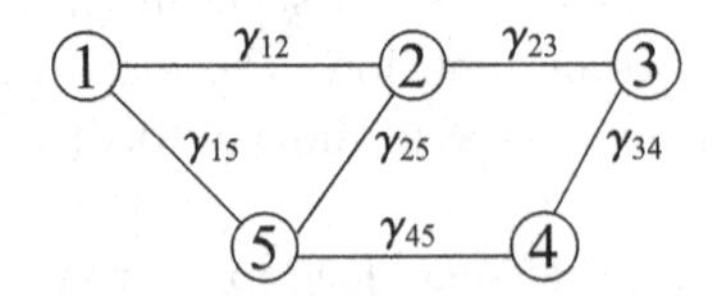

Figure 4.1 Random walk on a graph with transition rates γ_{ij}.

Clearly, the answer depends on the starting position, so let p_j be the expected reward starting from j. If $j \in S$ then, of course, $p_j = v_j$. The random walk with transition rates γ_{jk} is reversible, and its invariant distribution is uniform. By conditioning on the first vertex i to which the random walk jumps from j, we obtain the relations

$$p_j = \sum_i \frac{\gamma_{ji}}{\sum_k \gamma_{jk}} p_i, \qquad \text{for } j \in G \setminus S.$$

We can rewrite this set of equations as follows:

$$0 = \sum_i \overbrace{\underbrace{\gamma_{ij}}_{1/R} \underbrace{(p_i - p_j)}_{\Delta V}}^{\text{current}}, \qquad j \in G \setminus S,$$

$$p_j = v_j, \qquad j \in S.$$

Interpreting p_i as the voltage at vertex (node) i, and γ_{ij} as the *conductance* (inverse resistance) of the edge (ij), the first line is asserting that the sum of the currents through all the edges into a given node i is 0. These are Kirchhoff's equations for an electrical network G, in which nodes i and j are joined by a resistance of γ_{ij}^{-1}, and nodes $j \in S$ are held at potential v_j.

Can we develop a better understanding of why Kirchhoff's equations for the flow of current appear in this game?

4.1.2 Button model

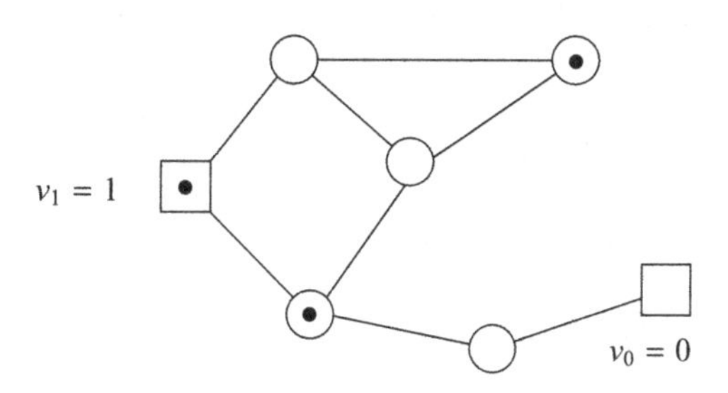

Figure 4.2 Button model.

Let us look at the special case of $S = \{0, 1\}$ with $v_0 = 0$ and $v_1 = 1$, and consider the following alternative model.

Suppose that on every node of the graph there is a button. Buttons on nodes j and k swap at rate γ_{jk}; i.e. associated with each edge there is an independent Poisson clock that ticks at rate γ_{jk}, and when it ticks the two buttons at the vertices of that edge swap places. When a button arrives at node 1, it is painted black; when it arrives at node 0, it is painted white. (A button may be repainted many times.) You should convince yourself that this defines a Markov process, whose state can be taken as the set of nodes where the buttons are coloured black.

As you will show in Exercise 4.1, each button is performing a symmetric random walk on the graph. Thus, for a given button, asking the question "am I black?" amounts to asking whether that button's individual random walk has more recently been to 1 or to 0. Since the random walk is reversible, the probability that the button is black is equal to the probability that the random walk started from that point will in the future visit node 1 before visiting node 0. Thus, from the point of view of this button, it is playing the game we described above, with $S = \{0, 1\}$.

Let us look at another equivalent model for this Markov process. Suppose that electrons perform a random walk on the graph with exclusion. That is, an electron in node i attempts to jump to a neighbouring node j at rate γ_{ij}, but, if there is already an electron in j, that jump will be blocked, and the state will be unchanged. Suppose also that electrons are pushed in at node 1, i.e., as soon as an electron leaves node 1, another one appears there; and electrons are pulled out at node 0, i.e., as soon as an electron appears there, it is removed from the system. (You will show the equivalence of the two models in Exercise 4.2.) This is a rather simple model of electron movement in an electrical network, with node 1 held at a higher voltage than node 0.

Let p_j be the (stationary) probability that node j is occupied by an electron. Then

$$p_0 = 0; \qquad p_1 = 1; \qquad p_j = \sum_i \frac{\gamma_{ji}}{\sum_k \gamma_{jk}} p_i, \quad j \neq 0, 1.$$

Equivalently,

$$0 = \sum_i \gamma_{ij}(p_i - p_j), \qquad j \neq 0, 1,$$

$$p_j = v_j, \qquad j = 0, 1,$$

which are precisely Kirchhoff's equations for an electrical network where nodes i and j are joined by a resistance γ_{ij}^{-1}, and nodes 0, 1 are held at a voltage of $v_0 = 0, v_1 = 1$.

The *net* flow of electrons from node j to node k is

$$\gamma_{jk}\mathbb{P}(j \text{ occupied}, k \text{ empty}) - \gamma_{kj}\mathbb{P}(k \text{ occupied}, j \text{ empty}).$$

Now, $\gamma_{jk} = \gamma_{kj}$. Also, we can rewrite the difference of probabilities by adding and subtracting the event that j and k are both occupied, as follows:

$$\begin{aligned}
&\mathbb{P}(j \text{ occupied}, k \text{ empty}) - \mathbb{P}(k \text{ occupied}, j \text{ empty}) \\
&\quad = (\mathbb{P}(j \text{ occupied}, k \text{ empty}) + \mathbb{P}(j \text{ occupied}, k \text{ occupied})) \\
&\quad\quad - (\mathbb{P}(j \text{ occupied}, k \text{ occupied}) + \mathbb{P}(k \text{ occupied}, j \text{ empty})) \\
&\quad = p_j - p_k.
\end{aligned}$$

Therefore, the net flow of electrons from node j to node k is

$$\gamma_{jk}(p_j - p_k),$$

i.e. we have recovered Ohm's law that the current flow from j to k is proportional to the voltage difference; the constant of proportionality is the conductance (inverse resistance) γ_{jk}.

We could also prove a result stating that, during a long time interval, the net number of electrons that have moved from j to k will be proportional to this quantity, and even give a central limit theorem for the "shot noise" around the average rate of electron motion from j to k.

4.1.3 Extremal characterizations

Next we give another angle of attack on current flow in networks, which was developed in the late nineteenth century.

Let u_{jk} be the current flowing from j to k, and let $r_{jk} = (1/\gamma_{jk})$ be the resistance between nodes j and k. Then the heat dissipation in the network is

$$\frac{1}{2}\sum_j \sum_k r_{jk} u_{jk}^2$$

(the factor of $1/2$ is there because we are double-counting each edge). Suppose we want to minimize the heat dissipation subject to a given total current U flowing from node 0 to node 1, and we are free to choose where current flows, subject only to flow balance at the other nodes. The problem is thus as follows:

$$\begin{array}{lll} \text{minimize} & \frac{1}{2}\sum_j \sum_k r_{jk}u_{jk}^2 & \\ \text{subject to} & \sum_k u_{jk} = \begin{cases} 0, & j \in G, \\ -U, & j = 0, \\ +U, & j = 1, \end{cases} & \\ \text{over} & u_{jk} = -u_{kj}, & j, k \in G. \end{array}$$

We can write the Lagrangian as

$$L(u; p) = \frac{1}{2}\sum_{j<k} r_{jk}u_{jk}^2 + \sum_j p_j \left(\sum_k u_{jk}\right) + p_0 U - p_1 U;$$

we shall see that the notation p_j for the Lagrange multipliers is not accidental. To deal with the condition $u_{jk} = -u_{kj}$, we simply eliminate u_{jk} with $j > k$, so that the Lagrangian involves only u_{jk} with $j < k$. (In particular, we read u_{kj} for $k > j$ as $-u_{jk}$ in the second term. Also, the summation in the first term is over $j < k$ rather than over all j and k; this corresponds to dividing the original objective function by 2.) Differentiating with respect to u_{jk},

$$\frac{\partial L}{\partial u_{jk}} = r_{jk}u_{jk} + p_j - p_k,$$

so the solution is

$$u_{jk} = \frac{p_k - p_j}{r_{jk}}.$$

That is, the Lagrange multipliers really are potentials in Kirchhoff's equations, and the currents u_{jk} obey Ohm's law with these potentials.

This is known as *Thomson's principle*: the flow pattern of current within a network of resistors is such as to minimize the heat dissipation for a given total current. (The Thomson in question is the physicist William Thomson, later Lord Kelvin.)

Extremal characterizations are very useful in deriving certain monotonicity properties of current flows. For example, suppose that we remove an edge from the network (i.e. assign an infinite resistance to it). What will this do to the effective resistance of the network? Intuitively, it seems clear that the effective resistance should increase, but how can we prove it?

This becomes an easy problem if we have found an extremal characterization. The effective resistance is the ratio of the heat dissipated to the square of the current, and so we will be done if we can show that removing an edge increases the minimum attained in the optimization problem. But removing an edge from the graph restricts the set of possible solutions, and so the minimal heat dissipation can only increase (or not go down, anyway) if we enforce $u_{jk} = 0$. It is quite tricky to prove this result without using the extremal characterization.

If we evaluate the heat dissipation in the network in terms of potential differences rather than currents, we are led to a dual form of the optimization problem:

$$\begin{array}{ll} \text{minimize} & \frac{1}{2}\sum_j \sum_k \gamma_{jk}(p_j - p_k)^2 \\ \text{subject to} & p_0 = 0, \ p_1 = 1, \\ \text{over} & p_j, \qquad j \in G. \end{array}$$

This yields the optimality conditions

$$\frac{\partial L}{\partial p_j} = \sum_k \gamma_{jk}(p_j - p_k) = 0, \qquad j \neq 0, 1,$$

which are again Kirchhoff's equations.

You may recognize the problem of minimizing a quadratic form subject to boundary conditions as the Dirichlet problem. (To see the connection with the classical Dirichlet problem on continuous spaces, imagine a network that is a square grid, and allow the grid to become finer and finer.)

Remark 4.1 It is interesting to note the parallels between the loss network of Chapter 3 and our model of an electrical network. At the microscopic level, we have a probabilistic model of call arrivals as a Poisson process and rules for whether a call is accepted; this parallels our description of electron motion as a random walk with exclusion. At the macroscopic level, we have quantities describing average behaviour over a period of time, such as blocking probabilities or currents and potentials, with relationships determining them, such as the conditions on B or Ohm's law and Kirchhoff's equations. Finally, at what might be termed the *teleological* level, we have for both networks an extremal characterization – an objective function that the network is "trying" to optimize.

We saw, in Figures 3.12 and 3.13, that instability of alternative routing could be interpreted in terms of multiple minima of a function. Viewed in this light, the bistability is a natural consequence of the form of the function that the network is trying to minimize. One way of interpreting the trunk reservation, sticky random scheme is as an attempt to align more closely the microscopic rules with the desired macroscopic consequences. (For example, in Exercise 3.10 you showed that the proportion of traffic rerouted from $i \to j$ via node k is inversely proportional to the loss rate on this route.)

Exercises

Exercise 4.1 Convince yourself that, if we look only at a single button in the button game, it is performing a symmetric random walk on a graph with transition rates γ_{jk}. (Of course, the random walks of different buttons will not be independent.)

Exercise 4.2 Convince yourself that, if we identify "electron" = "black button" and "no electron" = "white button", the button model is equivalent to the model of electron motion, in the sense that the set of nodes occupied by a black button and the set of nodes occupied by an electron both define the same Markov process. Why does swapping two black buttons correspond to a blocked electron jump?

Exercise 4.3 Show that the effective resistance of a network of resistors is not decreased if the resistance of an edge is increased.

4.2 Road traffic models

In Section 4.1 we saw that simple local rules may allow a network to solve a large-scale optimization problem. This is an encouraging insight; but local rules may lead to poor system behaviour if the rules are the wrong ones. We begin with an example that arises in the study of road traffic.

4.2.1 Braess's paradox

Figure 4.3 depicts a road network in which cars travel from south (S) to north (N). The labels on the edges, or one-way roads, indicate the delay that will be incurred by the cars travelling on that road, as a function of the traffic y (number of cars per unit time) travelling along it.

Let us fix the total flow from S to N at six cars per unit time. Figure 4.4(a) shows how the cars will distribute themselves. Note that all routes from S to N have the same total delay of 83 time units, so no driver has an incentive to switch route.

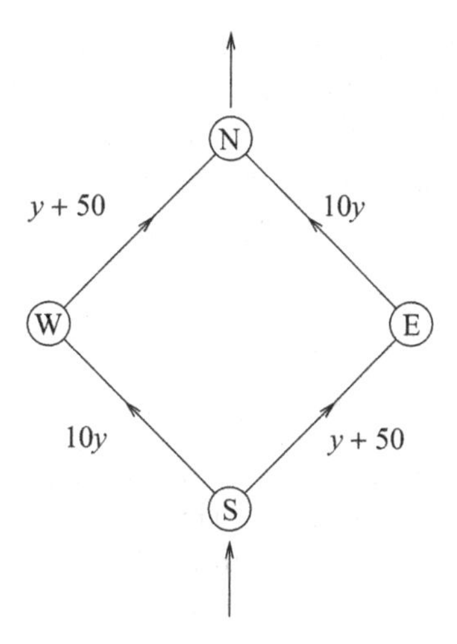

Figure 4.3 A road network with delays.

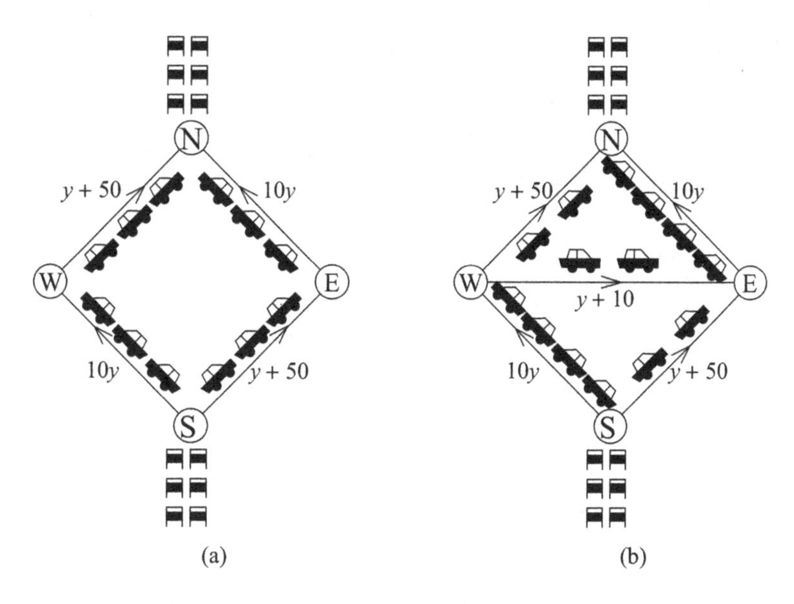

Figure 4.4 Braess's paradox. (a) Equilibrium flow from S to N. (b) Equilibrium when a new road is added.

In Figure 4.4(b), we have introduced an extra road with its own delay function, and found the new distribution of traffic such that no driver has an incentive to switch route. Note that all routes again have the same total delay, but it is now 92 time units! That is, adding extra road capacity has *increased* everyone's delay. This is in sharp contrast to electrical networks, where adding a link could only make it *easier* to traverse the network.

Next we define and analyze a general model of a road traffic network, with a view to figuring out why the paradox occurs and how we might avoid or fix it.

4.2.2 Wardrop equilibrium

We model the network as a set $\mathcal{J}$ of directed links. (If we want some of the roads to be two-way, we can introduce two links, one going in each direction.) The set of possible *routes* through the network is $\mathcal{R} \subset 2^{\mathcal{J}}$; each route is a subset of links (we do not need to know in what order these are traversed). Let A be the *link-route incidence matrix*, so $A_{jr} = 1$ if $j \in r$, and $A_{jr} = 0$ otherwise.

Let x_r be the flow on route r, and let $x = (x_r, r \in \mathcal{R})$ be the vector of flows. Then the flow on link j is given by

$$y_j = \sum_{r \in \mathcal{R}} A_{jr} x_r, \qquad j \in \mathcal{J}.$$

We can equivalently write $y = Ax$.

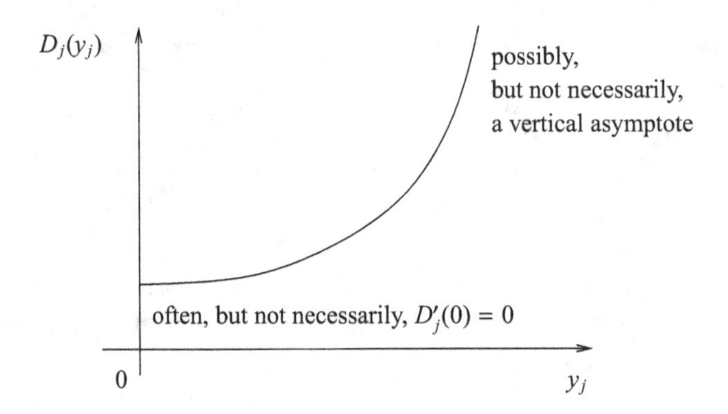

Figure 4.5 Possible delay as a function of traffic on a link.

The delay that is incurred on a single link j is given by a function $D_j(y_j)$, which we assume to be continuously differentiable and increasing. (We might also expect it to be convex, but we will not be using this assumption.) We will treat this delay as a steady-state quantity; it does not build up from one time period to another. The end-to-end delay along a route will simply be the sum of the delays along each link.

The basic premise of our model is that the individual cars care about getting from their origin to their destination, but don't care about the route they take. Let S be the set of source–destination pairs. For us, a source–destination pair is simply a set of routes that serve it. We let H be the incidence matrix with $H_{sr} = 1$ if the source–destination pair s is served by route r, and $H_{sr} = 0$ otherwise. (Column sums of H are 1, i.e. each route has a single source–destination pair that it serves. Let $s(r)$ be that source–destination pair for route r.) The flow f_s on a source–destination pair s is the sum of the flows along all the routes serving it:

$$f_s = \sum_{r \in \mathcal{R}} H_{sr} x_r, \qquad s \in S.$$

Equivalently, we write $f = Hx$.

We would like to answer the following question. Does there always exist a stable routing pattern, where none of the drivers has any incentive to switch routes? If so, can we characterize this routing pattern in a way that provides some insight into the paradox?

To illustrate the concepts of the link-route and route-destination incidence matrices, consider the network in Figure 4.6.

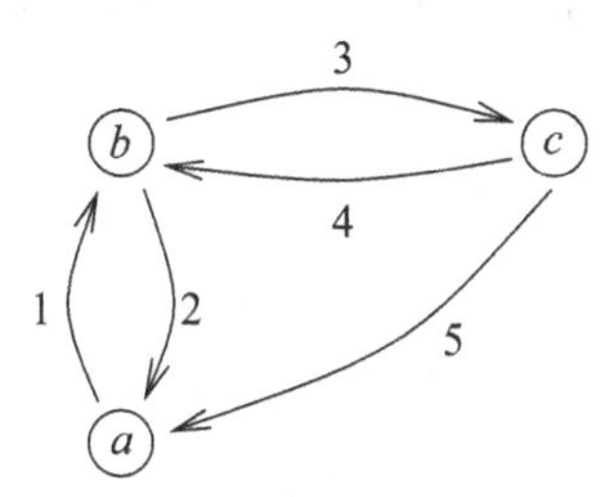

Figure 4.6 Links on an example network. Routes are $ab = \{1\}$, $ac = \{1, 3\}$, $ba = \{2\}$, $bc = \{3\}$, $ca_1 = \{5\}$, $ca_2 = \{4, 2\}$, $cb_1 = \{4\}$, $cb_2 = \{5, 1\}$.

We will take it to have the link-route incidence matrix

$$A = \begin{array}{c} \\ 1 \\ 2 \\ 3 \\ 4 \\ 5 \end{array} \begin{array}{c} \begin{array}{cccccccc} ab & ac & ba & bc & ca_1 & ca_2 & cb_1 & cb_2 \end{array} \\ \begin{pmatrix} 1 & 1 & 0 & 0 & 0 & 0 & 0 & 1 \\ 0 & 0 & 1 & 0 & 0 & 1 & 0 & 0 \\ 0 & 1 & 0 & 1 & 0 & 0 & 0 & 0 \\ 0 & 0 & 0 & 0 & 0 & 1 & 1 & 0 \\ 0 & 0 & 0 & 0 & 1 & 0 & 0 & 1 \end{pmatrix} \end{array}$$

and the corresponding route-destination incidence matrix

$$H = \begin{array}{c} \\ ab \\ ac \\ ba \\ bc \\ ca \\ cb \end{array} \begin{array}{c} \begin{array}{cccccccc} ab & ac & ba & bc & ca_1 & ca_2 & cb_1 & cb_2 \end{array} \\ \begin{pmatrix} 1 & 0 & 0 & 0 & 0 & 0 & 0 & 0 \\ 0 & 1 & 0 & 0 & 0 & 0 & 0 & 0 \\ 0 & 0 & 1 & 0 & 0 & 0 & 0 & 0 \\ 0 & 0 & 0 & 1 & 0 & 0 & 0 & 0 \\ 0 & 0 & 0 & 0 & 1 & 1 & 0 & 0 \\ 0 & 0 & 0 & 0 & 0 & 0 & 1 & 1 \end{pmatrix} \end{array}.$$

Note that we are omitting the potential two-link route serving *ba* and consisting of links 3 and 5; our modelling assumptions allow us to consider any set of routes, not necessarily all the physically possible ones.

Let us address what we mean by a stable routing pattern. Given a set of flows x_r, we can determine the delay of a driver on each of the routes. To find the delay of a driver on route r, we find the traffic y_j along each link of the network, evaluate the associated delay $D_j(y_j)$, and add up the delays along all the links of a given route. A routing pattern will be *stable* if none of the drivers has any incentive to switch: that is, for any route r serving a given source–destination pair and carrying a positive amount of traffic, the aggregate delay along all other routes serving the same source–destination pair is at least as large. Put into our notation, this becomes

$$x_r > 0 \implies \sum_{j \in \mathcal{J}} D_j(y_j) A_{jr} \leq \sum_{j \in \mathcal{J}} D_j(y_j) A_{jr'}, \qquad \forall\, r' \in s(r),$$

where recall that $s(r)$ is the set of routes serving the same source–destination pair as r. Such a stable pattern has a name.

Definition 4.2 A *Wardrop equilibrium* is a vector of flows along routes $x = (x_r, r \in \mathcal{R})$ such that

$$x_r > 0 \implies \sum_{j \in \mathcal{J}} D_j(y_j) A_{jr} = \min_{r' \in s(r)} \sum_{j \in \mathcal{J}} D_j(y_j) A_{jr'},$$

where $y = Ax$.

From our definition, it is not clear that a Wardrop equilibrium exists and how many of them there are. We will now show the existence of a Wardrop equilibrium, by exhibiting an alternative characterization of it.

Theorem 4.3 *A Wardrop equilibrium exists.*

Proof Consider the optimization problem

$$\begin{aligned} &\text{minimize} && \sum_{j\in\mathcal{J}} \int_0^{y_j} D_j(u)\,du \\ &\text{subject to} && Hx = f,\ Ax = y, \\ &\text{over} && x \geq 0,\ y. \end{aligned}$$

(Note that we are leaving y unconstrained; we will obtain $y \geq 0$ automatically, because $y = Ax$.)

The feasible region is convex and compact, and the objective function is differentiable and convex (because D_j is an increasing, continuous function). Thus, an optimum exists and can be found by Lagrangian techniques. The Lagrangian for the problem is

$$L(x, y; \lambda, \mu) = \sum_{j\in\mathcal{J}} \int_0^{y_j} D_j(u)\,du + \lambda\cdot(f - Hx) - \mu\cdot(y - Ax).$$

To minimize, we differentiate:

$$\frac{\partial L}{\partial y_j} = D_j(y_j) - \mu_j, \qquad \frac{\partial L}{\partial x_r} = -\lambda_{s(r)} + \sum_j \mu_j A_{jr}.$$

We want to find a minimum over $x_r \geq 0$ and all $y_j \in \mathbb{R}$. This means that at the minimum the derivative with respect to y_j is equal to 0, i.e. we can identify $\mu_j = D_j(y_j)$ as the delay on link j. The derivative with respect to x_r must be non-negative, and 0 if $x_r > 0$, so

$$\lambda_{s(r)} \begin{cases} = \sum_j \mu_j A_{jr}, & x_r > 0, \\ \leq \sum_j, \mu_j A_{jr}, & x_r = 0. \end{cases}$$

Therefore, we can interpret $\lambda_{s(r)}$ as the minimal delay available to the source–destination pair $s(r)$.

Consequently, solutions of this optimization problem are in one-to-one correspondence with Wardrop equilibria. □

Remark 4.4 Can we interpret the function $\int_0^{y_j} D_j(u)\,du$ appearing in the above optimization? Not that easily: the glib answer is that it is the function whose derivative is $D_j(y_j)$. The reader with a well-developed physics insight might recall the relationship between potential energy and force: if the force acting on an object is a function of position only, then the object's potential energy is the function whose derivative is the (negative of the) force. Despite this indirect definition, we have a well-developed intuition about energy, and of the tendency of physical systems to move to minimum potential energy configurations.

Later, in Exercise 7.13, we touch on an economic parallel in which an abstract concept, utility, is defined as a function whose derivative gives a measurable quantity, demand.

Remark 4.5 Is the Wardrop equilibrium unique? Strictly speaking, no. If we assume that all D_j are strictly increasing, we can conclude that there is a unique optimum for the link flows y in the optimization problem above, since the objective function will be a strictly convex function of y. However, for a general link-route incidence matrix, there is no reason to expect uniqueness in x. For example, in the network in Figure 4.7, we can clearly shift traffic between the solid and the dashed routes while keeping the traffic on each link the same.

Of course, as in Chapter 3, if we assume the existence of single-link routes, the Wardrop equilibrium will be unique, but it is somewhat less natural to assume the existence of single-link traffic here.

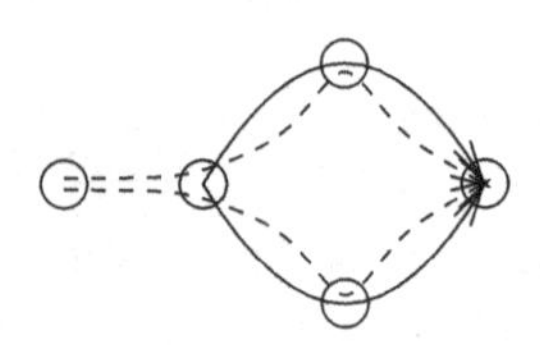

Figure 4.7 Network with four links: two two-link routes (solid) and two three-link routes (dashed). This network has many traffic patterns x giving the same link flows y.

In the course of showing the existence of the Wardrop equilibrium, we have arrived at a certain extremal characterization of it: it is a set of traffic flows that optimizes, subject to constraints, the quantity $\sum_{j\in\mathcal{J}} \int_0^{y_j} D_j(u)\,du$. If we add an extra link (and, therefore, extra possible routing options), this quantity will decrease. However, this tells us nothing about the changes to the average delay.

How might we design a network so that the choices of users would line up with a societal goal of minimizing the average delay? Consider the earlier problem but with a different objective function:

$$\begin{array}{ll} \text{minimize} & \sum_{j \in \mathcal{J}} y_j D_j(y_j) \\ \text{subject to} & Hx = f, \ Ax = y, \\ \text{over} & x \geq 0, \ y. \end{array}$$

The quantity $\sum_j y_j D_j(y_j)$ is the rate at which the cumulative total delay of users of the system increases. (In our model, which has a constant flow of users through the system per unit time, this is the natural quantity to consider.) Applying the same techniques to this new minimization problem, we get the Lagrangian

$$L(x, y; \lambda, \mu) = \sum_{j \in \mathcal{J}} y_j D_j(y_j) + \lambda \cdot (f - Hx) - \mu \cdot (y - Ax).$$

When we differentiate it, we obtain

$$\frac{\partial L}{\partial y_j} = D_j(y_j) + y_j D_j'(y_j) - \mu_j, \qquad \frac{\partial L}{\partial x_r} = -\lambda_{s(r)} + \sum_j \mu_j A_{jr}.$$

At the optimum, we still have

$$\lambda_{s(r)} \begin{cases} = \sum_j \mu_j A_{jr}, & x_r > 0, \\ \leq \sum_j \mu_j A_{jr}, & x_r = 0, \end{cases}$$

i.e. λ is the minimal total "something" along the route. The quantity μ_j, however, now has an extra term:

$$\mu_j = D_j(y_j) + y_j D_j'(y_j).$$

If we interpret $T_j(y_j) = y_j D_j'(y_j)$ as a congestion toll that the users of link j must pay, then μ_j is their total cost from both the delay and the toll, and $\lambda_{s(r)}$ is the minimal cost available to source–destination pair $s(r)$.

This suggests that, by adding tolls on the links, we may be able to encourage users towards collectively more desirable behaviour. It is, of course, still a very simple model of a complex social problem. For example, it assumes that delay and toll are weighted equally by each driver. But the optimization formulations do give us insight into the original paradox. The pattern of traffic in Figure 4.7 is optimizing a certain function, but it is not the "right" one. So it should not surprise us that the solution has counterintuitive properties.

Exercises

Exercise 4.4 Calculate the toll $T_j(y) = yD'_j(y)$ on each of the nine links in Figure 4.4. Observe that, if this toll is applied just after the new road is built (i.e. while nobody is using it), there is no incentive for any driver to use the new road.

Exercise 4.5 In the definition of a Wardrop equilibrium, f_s is the aggregate flow for source–sink pair s, and is assumed fixed. Suppose now that the aggregate flow between source–sink pair s is not fixed, but is a continuous, strictly decreasing function $B_s(\lambda_s)$, where λ_s is the minimal delay over all routes serving the source–sink pair s, for each $s \in \mathcal{S}$. For the extended model, show that an equilibrium exists and solves the optimization problem

$$\begin{aligned} \text{minimize} \quad & \sum_{j \in \mathcal{J}} \int_0^{y_j} D_j(u)\,du - G(f) \\ \text{subject to} \quad & Hx = f, Ax = y, \\ \text{over} \quad & x \geq 0, y, f, \end{aligned}$$

for a suitable choice of the function $G(f)$. Interpret your result in terms of a fictitious additional one-link "stay-at-home" route for each source–destination pair s, with appropriate delay characteristics.
[*Hint:* $G(f) = \sum_{s \in \mathcal{S}} \int^{f_s} B_s^{-1}(u)\,du$.]

Exercise 4.6 Recall the Dual optimization problem for loss networks:

$$\begin{aligned} \text{minimize} \quad & \underbrace{\sum_r \nu_r e^{-\sum_j y_j A_{jr}}}_{\text{carried traffic}} + \sum_j y_j C_j \\ \text{subject to} \quad & y \geq 0. \end{aligned}$$

Thus the function being minimized does not readily align with, for example, the sum of carried traffic, which we might want to *maximize*. Consider a three-link loss network with

$$A = \begin{pmatrix} 1 & 0 & 0 & 1 \\ 0 & 1 & 0 & 1 \\ 0 & 0 & 1 & 1 \end{pmatrix},$$

with $\nu_j = C_j = 100$, $j = 1, 2, 3$. Show that if ν_{123} is large enough, reducing the capacity of link 1 to zero will *increase* the carried traffic.

4.3 Optimization of queueing and loss networks

Models similar to those of Section 4.2 have been used for traffic in communication networks: in this section we consider queueing networks, where the ideas cross over easily, and also loss networks.

4.3.1 Queueing networks

Consider an open network of queues. As in the previous models, we assume that there are routes through this network; a route is simply a set of queues that the customer must traverse. Let ϕ_j be the service rate at queue j, and let ν_r be the arrival rate of customers on route r. We shall suppose the mean sojourn time of customers in queue j is $1/(\phi_j - \lambda_j)$, where $\lambda_j = \sum_{r:j\in r} \nu_r$. (This model would arise from a network of $\cdot/M/1$ queues, an example of a network with a product-form stationary distribution.)

Suppose that a unit delay of a customer on route r costs w_r; then the mean cost per unit time in the system is given by

$$W(\nu;\phi) = \sum_r w_r \sum_{j:j\in r} \frac{\nu_r}{\phi_j - \sum_{r':j\in r'} \nu_{r'}},$$

where $\nu = (\nu_r)_r$, $\phi = (\phi_j)_j$. There may be multiple routes r that serve the same source–destination traffic, and we may be able to choose how we divide the traffic across these routes. Can we tell if it would be desirable to shift traffic from one such route to another?

Now consider increasing the arrival rate on route r slightly. The change in total delay will be given by

$$\frac{dW}{d\nu_r} = \sum_{j\in r}\Bigg(\underbrace{\frac{w_r}{\phi_j - \lambda_j}}_{\substack{\text{extra delay at queue } j \\ \text{of another customer on route } r}} + \underbrace{\sum_{r':j\in r'} \frac{\nu_{r'} w_{r'}}{(\phi_j - \lambda_j)^2}}_{\substack{\text{knock-on cost, or externality,} \\ \text{to later customers on routes through } j \\ \text{of another customer on route } r}} \Bigg).$$

The expression gives the exact derivative, and the annotation below the expression gives an intuitively appealing interpretation: a single extra customer on route r will incur delay itself at each queue along its route, and it will additionally cause a knock-on cost to other customers passing through these queues.

If two or more routes can substitute for each other, the objective function $W(\nu;\phi)$ can be improved by shifting traffic towards routes with smaller derivatives. (Gallager (1977) describes an algorithm to implement this via routing rules held at individual nodes of a data network.)

4.3.2 Implied costs in loss networks

What happens if we explicitly attempt to align routing in a loss network, so as to optimize a weighted sum of carried traffics? In this section, we use the Erlang fixed point to conduct an analysis for a loss network similar to that for the queueing network of Section 4.3.1. The calculations are more involved for a loss network, since an increase in traffic on a route affects not just the resources on that route: the knock-on effects spread out throughout the network. Nevertheless, we shall see that the equations we derive still have a local character.

Consider then the fixed point model described in Chapter 3. Let the blocking probabilities $B_1, B_2, \ldots, B_J$ be a solution to the equations

$$B_j = E(\rho_j, C_j), \qquad j = 1, 2, \ldots, J, \tag{4.1}$$

where

$$\rho_j = (1 - B_j)^{-1} \sum_r A_{jr} \nu_r \prod_i (1 - B_i)^{A_{ir}}, \tag{4.2}$$

and the function E is Erlang's formula,

$$E(\nu, C) = \frac{\nu^C}{C!} \left[\sum_{n=0}^{C} \frac{\nu^n}{n!} \right]^{-1}. \tag{4.3}$$

Then an approximation for the loss probability on route r is given by

$$1 - L_r = \prod_j (1 - B_j)^{A_{jr}}. \tag{4.4}$$

Suppose that each call carried on route r is worth w_r. Then, under the approximation (4.4), the rate of return from the network will be

$$W(\nu; C) = \sum_r w_r \lambda_r, \qquad \text{where} \quad \lambda_r = \nu_r(1 - L_r)$$

corresponds to the traffic carried on route r. We emphasize the dependence of W on the vectors of offered traffics $\nu = (\nu_r, r \in \mathcal{R})$ and capacities $C = (C_1, C_2, \ldots, C_J)$, and we shall be interested in how $W(\nu; C)$ varies with changes in traffic on different routes, or with changes in capacity at different links. Erlang's formula can be extended to non-integral values of its second argument in various ways, but for our purpose there is definitely a preference: we extend the definition (4.3) to non-integral values of the scalar C by linear interpolation, and at integer values of C_j we define the derivative of $W(\nu; C)$ with respect to C_j to be the left derivative. (These definitions are explored in Exercise 4.8.)

Let

$$\delta_j = \rho_j(E(\rho_j, C_j - 1) - E(\rho_j, C_j)) .$$

(This is *Erlang's improvement formula* – the increase in carried traffic that comes from a unit increase in capacity.)

Theorem 4.6

$$\frac{d}{dv_r} W(v; C) = (1 - L_r)s_r \tag{4.5}$$

and

$$\frac{d}{dC_j} W(v; C) = c_j, \tag{4.6}$$

where $s = (s_r, r \in \mathcal{R})$ *and* $c = (c_1, c_2, \ldots, c_J)$ *are the unique solution to the linear equations*

$$s_r = w_r - \sum_j c_j A_{jr}, \tag{4.7}$$

$$c_j = \delta_j \frac{\sum_r A_{jr}\lambda_r(s_r + c_j)}{\sum_r A_{jr}\lambda_r}. \tag{4.8}$$

Remark 4.7 We can interpret s_r as the *surplus value* of a call on route r: if such a call is accepted, it will earn w_r directly, but at an *implied cost* of c_j for each circuit used from link j. The implied costs c measure the expected knock-on effects of accepting a call upon later arrivals at the network. From (4.6) it follows that c_j is also a *shadow price*, measuring the sensitivity of the rate of return to the capacity C_j of link j. Note the local character of equations (4.7) and (4.8): the right-hand side of (4.7) involves costs c_j only for links j on the route r, while (4.8) exhibits c_j in terms of a weighted average, over just those routes through link j, of $s_r + c_j$. We can interpret (4.8) as follows: a circuit used for a call on link j will cause the loss of an additional arriving call with probability δ_j; given that it does, the probability the additional lost call is on route r is proportional to $A_{jr}\lambda_r$; and this will lose w_r directly, but have a positive knock-on impact elsewhere on route r, captured by the term $s_r + c_j$.

Proof Note that in many ways this is not a complicated statement, as all we are doing is differentiating an implicit function. However, we are going to be somewhat clever about it.

We shall prove the theorem for the case where (A_{jr}) is a 0–1 matrix. Suppose, without loss of generality, that there exist marker routes $\{j\} \in \mathcal{R}$

for $j = 1, 2, \ldots, J$, with $\nu_{\{j\}} = w_{\{j\}} = 0$. For notational simplicity, we will simply write ν_j for $\nu_{\{j\}}$.

Exercise 4.8 shows that, in order to compute dW/dC_j, it is sufficient to compute $dW/d\nu_j$ instead. The key observation is that if, when we change ν_j, we also appropriately change the values of all the other ν_k, we can keep the loads ρ_k and blocking probabilities B_k on all links $k \neq j$ unchanged. This makes it easy to compute the resulting change in W. (To avoid trivial complications with what it means to have $\nu_j < 0$, we can compute the derivatives around some point where all ν_j are positive; the algebraic machinery will be exactly the same.)

We proceed as follows. Alter the offered traffic ν_j. This will affect directly the blocking probability B_j at link j, and hence the carried traffics λ_r for routes r through link j. This in turn will have indirect effects upon other links through which these routes pass. We can, however, cancel out these indirect effects by judicious alterations to ν_k for $k \neq j$. The alterations to ν_k have to be such as to leave the reduced load ρ_k constant for $k \neq j$, as then, from (4.1), the blocking probability B_k will be left constant for $k \neq j$.

Let us begin by calculating the direct effect of the change in ν_j on the carried traffic λ_r, along a route through link j. From the relation

$$\lambda_r = \nu_r \prod_k (1 - B_k)^{A_{kr}},$$

assuming we can arrange for all B_k, $k \neq j$, to be unchanged, the direct effect is

$$d\lambda_r = -A_{jr}(1 - B_j)^{-1}\lambda_r \cdot \frac{\partial B_j}{\partial \nu_j} \cdot d\nu_j.$$

Here the partial derivative $\partial B_j/\partial \nu_j$ is calculated by allowing ν_j to vary, but holding the other traffic offered to link j fixed. We can use Exercise 4.8 to relate δ_j to this partial derivative, and deduce that

$$\begin{aligned}\frac{\partial B_j}{\partial \nu_j} &= \frac{d}{d\rho_j} E(\rho_j, C_j) \\ &= (1 - B_j)\delta_j \rho_j^{-1}.\end{aligned}$$

Thus

$$d\lambda_r = -A_{jr}\lambda_r \delta_j \rho_j^{-1} \cdot d\nu_j.$$

Next we calculate the necessary alterations to ν_k for $k \neq j$. In order that ρ_k be left constant, for each route r passing through k we must compensate for the change $d\lambda_r$ by a change

$$dv_k = -A_{kr}(1 - B_k)^{-1} \cdot d\lambda_r.$$

Indeed, from (4.2), reduced load ρ_k receives a contribution $A_{kr}(1 - B_k)^{-1}\lambda_r$ from route r. Observe that, apart from marker routes, the only routes for which λ_r changes are routes through link j. Therefore, we can calculate the effect of this combination of changes of the v_j on $W(v; C)$, as we have computed $d\lambda_r$ for those routes already. Putting the various terms together, we obtain

$$\left[\frac{d}{dv_j} + \sum_r A_{jr}\lambda_r\delta_j\rho_j^{-1} \sum_{k \neq j} A_{kr}(1 - B_k)^{-1} \frac{d}{dv_k}\right] W(v; C) = -\sum_r A_{jr}\lambda_r\delta_j\rho_j^{-1} w_r \,. \tag{4.9}$$

Further, using (4.6) and Exercise 4.8, we obtain

$$c_j = -(1 - B_j)^{-1} \frac{d}{dv_j} W(v; C). \tag{4.10}$$

Substituting (4.10) into (4.9) gives (4.8), where we use (4.7) to define s_r.

To calculate the derivative (4.5), compensate for the change in v_r by alterations to v_j that hold ρ_j, and hence B_j, constant; that is,

$$dv_j = -A_{jr}(1 - B_j)^{-1} \prod_k (1 - B_k)^{A_{kr}} \cdot dv_r.$$

The change in carried traffic on route r, worth w_r, is $(1 - L_r)dv_r$. Thus

$$\left[\frac{d}{dv_r} - \sum_j A_{jr}(1 - B_j)^{-1} \prod_k (1 - B_k)^{A_{kr}} \frac{d}{dv_j}\right] W(v; C) = (1 - L_r)w_r \,.$$

The derivative (4.5) now follows from the identity (4.10).

Finally, the uniqueness of the solution to equations (4.7) and (4.8) is shown in Exercise 4.9. □

Remark 4.8 If traffic serving a source–sink pair can be spread over two or more routes, the derivatives we have calculated for queueing and loss networks can be used as the basis for a gradient descent algorithm to find a local optimum of the function W. This form of routing optimization is called *quasi-static*: it corresponds to probabilistically splitting traffic over routes according to proportions that vary slowly, so that the traffic offered to each route is approximately Poisson over short time periods. In contrast, *dynamic* schemes, such as the sticky random scheme of Section 3.7.2, route calls according to the instantaneous state of the network.

The discussion of resource pooling in Section 3.7.3 showed that we should in general expect dynamic schemes to achieve better performance than quasi-static schemes, as a consequence of trunking efficiency. But we saw, in the same discussion, that, in networks with efficient resource pooling, there may be little warning that the system as a whole is close to capacity. Implied costs can be calculated for dynamic routing schemes, where they can help anticipate the need for capacity expansion before it becomes apparent through loss of traffic (see Key (1988)). Together with measurements of load, these calculations can be very helpful with longer-term network planning, since externalities, such as those that caused the difficulties with Braess's paradox, are taken into account.

Exercises

Exercise 4.7 Show that the function $W(\nu;\phi)$ defined in Section 4.3.1 for a queueing network is a strictly convex function of ν, and deduce that a gradient descent algorithm will eventually find the minimum. Show that when w_r does not depend upon r, the queueing network can be recast in terms of the final optimization in Section 4.2, and check that the toll $T_j(y)$ defined there is just the externality identified in Section 4.3.1.

Exercise 4.8 For scalar ν, C recall from Exercise 1.8 that

$$\begin{aligned}\frac{d}{dC}E(\nu,C) &\equiv E(\nu,C) - E(\nu,C-1)\\ &= -(1-E(\nu,C))^{-1}\frac{d}{d\nu}E(\nu,C)\,.\end{aligned}$$

For vector ν, C show that

$$\frac{d}{dC_j}W(\nu;C) = -(1-B_j)^{-1}\frac{d}{d\nu_j}W(\nu;C).$$

Exercise 4.9 Show that equations (4.7) and (4.8) may be rewritten in an equivalent form as a matrix identity

$$c(I + A\lambda A^T\gamma) = w\lambda A^T\gamma,$$

where we define diagonal matrices

$$\lambda = \mathrm{diag}(\lambda_r, r\in\mathcal{R}),\quad \gamma = \mathrm{diag}\left(\frac{\delta_j}{(1-\delta_j)\sum_r A_{jr}\lambda_r},\ j\in\mathcal{J}\right).$$

Note that $(I + \gamma^{1/2} A\lambda A^T \gamma^{1/2})$ is positive definite and hence invertible. Deduce that the equations have the unique solution

$$c = w\lambda A^T \gamma^{1/2} (I + \gamma^{1/2} A\lambda A^T \gamma^{1/2})^{-1} \gamma^{1/2} .$$

4.4 Further reading

For the reader interested in the history of the subject, Erlang (1925) gives an insight into the influence of Maxwell's law (for the distribution of molecular velocities) on Erlang's thinking, and Doyle and Snell (2000) give a beautifully written account of work on random walks and electric networks going back to Lords Rayleigh and Kelvin. Whittle (2007) treats the optimization of network design in a wide range of areas, with fascinating insights into many of the topics covered here.

5

Random access networks

Consider the situation illustrated in Figure 5.1. We have multiple base stations which cannot talk to each other directly because the Earth is in the way. (The diagram is not to scale.) Instead, they send messages to a satellite. The satellite will broadcast the message (with its address) back to all the stations.

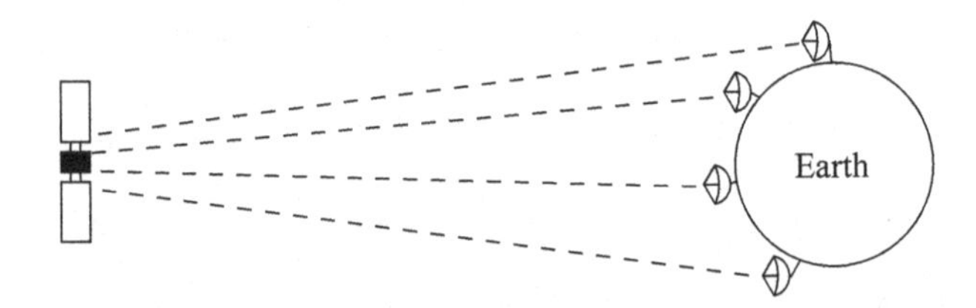

Figure 5.1 Multiple base stations contacting a satellite.

If two stations attempt to transmit messages simultaneously, the messages will interfere with each other, and will need to be retransmitted. So the fundamental question we will be addressing here is *contention resolution*, i.e. avoiding collisions in such a set-up.

Further examples of the same fundamental problem occur with mobile telephones attempting connection to a base station, or wireless devices communicating with each other. In all cases, if we could instantaneously sense whether someone else is transmitting, there would be no collisions. The problem arises because the finite speed of light causes a delay between the time when a station starts transmitting and the time when the other stations sense this interference. As processing speeds increase, the speed-of-light delays pose a problem over shorter and shorter distances.

One approach might be to use a *token ring*. Label the stations in a cyclical order. One by one, each station will transmit its messages to the satellite. When it is done, it will send a "token" message to indicate that it is finished; this lets the next station begin transmitting. If a station has no messages to transmit, it will simply pass on the token. This approach has difficulties if the token becomes corrupted, e.g. by ambient noise. But it also does not work well in an important limiting regime where the number of stations is large (and the total traffic is constant), since simply passing on the token around all the stations will take a long time.

Our focus in this chapter will be on protocols that are useful when the number of stations is large or unknown. These protocols use randomness in an intrinsic manner in order to schedule access to the channel, hence the title of the chapter.

5.1 The ALOHA protocol

The ALOHA protocol was developed for a radio broadcast network connecting terminals on the Hawaiian islands to a central computing facility, hence its name.

We assume that time is divided into discrete slots, and that all transmission attempts take up exactly one time slot. Let the channel state be $Z_t \in \{0, 1, *\}$, $t \in \mathbb{Z}_+$. We put $Z_t = 0$ if no transmission attempts were made during time slot t, $Z_t = 1$ if exactly one transmission attempt was made, and $Z_t = *$ if more than one transmission attempt was made.

We suppose that new packets for transmission arrive in a Poisson stream of rate ν, so that in each time slot the random number of newly arriving packets has a Poisson distribution with mean ν. We model each packet as corresponding to its own station; once the packet is successfully transmitted, the station leaves the system. Thus, there is no queueing of packets at a station in this model.

Let Y_t be the number of new packets arriving during the time slot $t - 1$. These are all at new stations, and we suppose each of them attempts transmission during time slot t. In addition, stations that previously failed to transmit their packets may attempt to retransmit them during the time slot. A transmission attempt during time slot t is successful only if $Z_t = 1$; if $Z_t = *$, there is a collision, and none of the packets involved are successfully transmitted in this slot.

We have so far not described the retransmission policy. We now define it as follows. Let $f \in (0, 1)$ be a fixed parameter.

Definition 5.1 (ALOHA protocol) After an unsuccessful transmission attempt, a station attempts retransmission after a delay that is a geometrically distributed random variable with mean f^{-1}, independently of everything else. Equivalently, after its first attempt, a station independently retransmits its packet with probability f in each following slot until it is successfully transmitted.

Let N_t be the backlog of packets awaiting retransmission, i.e. packets that arrived before time slot $t-1$ and have not yet been successfully transmitted. Then, under the ALOHA protocol, the number of packets that will attempt *re*transmission during the tth slot is binomial with parameters N_t and f. Consequently, the total number of transmission attempts during time slot t is

$$A_t = Y_t + \text{Binom}(N_t, f).$$

The channel state Z_t is 0, 1 or $*$ according to whether A_t is 0, 1 or >1. The backlog evolves as

$$N_{t+1} = N_t + Y_t - I\{Z_t = 1\},$$

and under our Poisson and geometric assumptions N_t is a Markov chain. We will try to determine whether it is recurrent. First, let's look at the *drift* of the backlog, i.e. the conditional expectation of its change:

$$\begin{aligned} \mathbb{E}[N_{t+1} - N_t \mid N_t = n] &= \mathbb{E}[Y_t - I\{Z_t = 1\} \mid N_t = n] \\ &= \nu - \mathbb{P}(Z_t = 1 \mid N_t = n), \end{aligned}$$

where the first term follows since Y_t is an independent Poisson variable with mean ν. To calculate the second term, note that $Z_t = 1$ can occur only if $Y_t = 1$ and there are no retransmission attempts, or if $Y_t = 0$ and there is exactly one retransmission attempt. If $Y_t > 1$, we are guaranteed that $Z_t = *$. Thus

$$\mathbb{P}(Z_t = 1 \mid N_t = n) = e^{-\nu} \cdot nf(1-f)^{n-1} + \nu e^{-\nu} \cdot (1-f)^n.$$

We conclude that the drift is positive (i.e. backlog is "on average" growing) if

$$\nu > e^{-\nu}(nf + (1-f)\nu)(1-f)^{n-1}.$$

For any fixed retransmission probability f, the quantity on the right-hand side tends to 0 as $n \to \infty$. Consequently, for any positive arrival rate of messages ν, if the backlog is large enough, we expect it to grow even larger. This strongly suggests that the backlog will be transient – not a good system feature!

Remark 5.2 The drift condition does not prove that N_t is transient: there are recurrent Markov chains with positive drift. For example, consider the

Markov chain where from each state n there are two possible transitions: to $n + 2$ with probability $1 - 1/n$, and to 0 with probability $1/n$. Then the drift is $1 - 2/n \to 1$, but $\mathbb{P}(\text{hit 0 eventually}) = 1$ so the chain is recurrent. (The drift condition *does* show that the chain cannot be positive recurrent.)

For our process, the jumps are constrained, and it is thus easy to show that $\mathbb{E}[(N_{t+1} - N_t)^2 \mid N_t]$ is uniformly bounded; this, together with the drift analysis, can lead to a proof of transience. We will not pursue this approach.

We shall see that the system has an even more clear failing than a growing backlog.

Proposition 5.3 *Consider the ALOHA protocol, with arrival rate $\nu > 0$ and retransmission probability $f \in (0, 1)$. Almost surely, there exists a finite (random) time after which we will always have $Z_t = *$. That is, ALOHA transmits only finitely many packets and then "jams" forever. Formally,*

$$\mathbb{P}(\exists J < \infty : Z_t = * \ \forall t \geq J) = 1.$$

Remark 5.4 The time J is not a stopping time. That is, given the history of the system up until time t, we cannot determine whether $J < t$ and the channel has jammed.

Proof For each size of the backlog, consider the probability that the channel will unjam before the backlog increases:

$$\begin{aligned} p(n) &= \mathbb{P}(\text{channel unjams before backlog increases} \mid N_0 = n) \\ &= \mathbb{P}(\exists T < \infty : N_1, \ldots, N_T = n, Z_T = 0 \text{ or } 1 \mid N_0 = n). \end{aligned}$$

To compute $p(n)$, let us think of this as a game. Suppose we reach time t without either the channel unjamming or the backlog increasing. If at time t, $Z_t = 0$ or 1, then $T = t$ and we "win". If $Z_t = *$ and $N_{t+1} > n$, then no such T exists and we "lose". Otherwise, $Z_t = *$ and $N_{t+1} = n$, and we are in the same situation at time $t + 1$; call this outcome "try again". In such a game, the probability of eventually winning is

$$\begin{aligned} p(n) &= \frac{\mathbb{P}(\text{win})}{\mathbb{P}(\text{win}) + \mathbb{P}(\text{lose})} = \frac{\mathbb{P}(Z_t = 0 \text{ or } 1 \mid N_t = n)}{1 - \mathbb{P}(N_{t+1} = n, Z_t = * \mid N_t = n)} \\ &= \frac{e^{-\nu}(1 + \nu)(1 - f)^n + e^{-\nu} n f (1 - f)^{n-1}}{1 - e^{-\nu}(1 - (1 - f)^n - n f (1 - f)^{n-1})} \\ &\sim \frac{n f (1 - f)^{n-1}}{e^{\nu} - 1} \text{ as } n \to \infty. \end{aligned}$$

As we expected, $p(n) \to 0$ as $n \to \infty$. Moreover, they are summable:

$$\sum_{n=0}^{\infty} p(n) < \infty.$$

Summable sequences of probabilities bring to mind the first Borel–Cantelli lemma.

Theorem 5.5 (First Borel–Cantelli lemma) *If (A_n) is a sequence of events such that $\sum_{n=1}^{\infty} \mathbb{P}(A_n) < \infty$, then the probability that infinitely many of the events occur is 0.*

Proof

$$\begin{aligned}\mathbb{E}[\text{number of events occurring}] &= \mathbb{E}\left[\sum_n I[A_n]\right] \\ &= \sum_n \mathbb{E}[I[A_n]] = \sum_n \mathbb{P}(A_n) < \infty.\end{aligned}$$

Since the number of events occurring has finite expectation, it must be finite with probability 1. □

We cannot apply the Borel–Cantelli lemma directly to the sequence $p(n)$: we need a sequence of events that occur once or not at all, whereas there may be many times when the backlog reaches level n. Instead, we will look at the *record values* hit at *record times*. Set $R(1) = 1$ and let $R(r+1) = \inf\{t > R(r) : N_t > N_{R(r)}\}$.

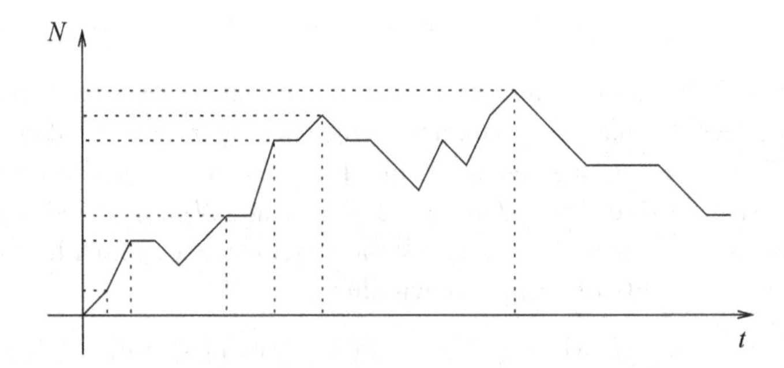

Figure 5.2 A possible trajectory of N_t, and its record values and record times.

By definition, we hit each record value only once. Moreover, with probability 1 the sequence of record values is infinite, i.e. the backlog is unbounded.

Finally, since we showed $\sum_{n=1}^{\infty} p(n) < \infty$, we certainly have

$$\sum_{r=1}^{\infty} p(N_{R(r)}) \leq \sum_{n=0}^{\infty} p(n) < \infty.$$

Let $A(r)$ be the event that the channel unjams after the backlog hits its rth record value. By the first Borel–Cantelli lemma, only finitely many of the events $A(r)$ occur. However, since we hit the rth record value at a finite time, this means that there will be only finitely many successful transmissions:

$$\mathbb{P}(\exists J < \infty : Z_t = * \ \forall t \geq J) = 1,$$

for any arrival rate $\nu > 0$. □

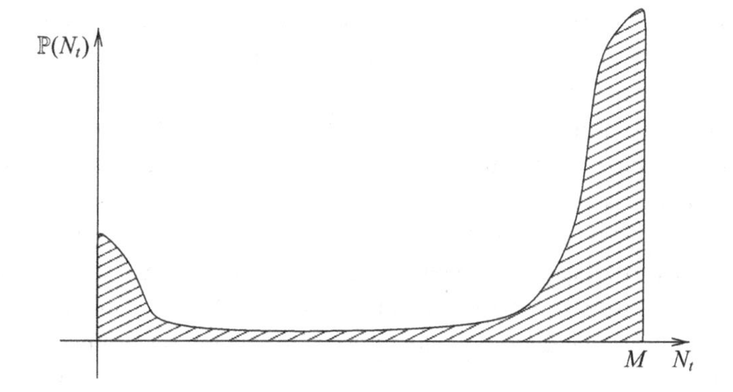

Figure 5.3 Qualitative behaviour of the equilibrium distribution of the backlog with a finite number of stations.

Remark 5.6 Suppose the number of stations is large but finite, say M, and that stations that have a backlog do not generate new packets. In that case, the natural assumption is that the number of new arrivals in a time slot has a binomial distribution $Y_t \sim \text{Binom}(M - N_t, q)$ rather than a Poisson distribution.

Since N_t has a state space that is finite and irreducible, it can't be transient. But Mf may still be large, in which case the equilibrium distribution of the backlog has the form sketched in Figure 5.3. While the backlog is small, it tends to stay small. However, if the backlog grows large at some point, it will tend to stay large for a really long time. This bistability for finite M corresponds to transience in the Poisson limit ($M \to \infty$ with $Mq = \nu$ held fixed).

It may take a very long time for the system to traverse from the left to the right of Figure 5.3 (see the discussion of bistability in Section 3.7.1), and the theoretical instability may not be a real problem for many systems. Nevertheless, motivated by large-scale systems where there may be no apparent upper bound on the number of stations, the theoretical question arises: is it possible to construct a random access scheme that has a stable throughput (i.e. a long-run proportion of successfully occupied slots that is positive) for some rate $\nu > 0$? We address this in the following two sections. And we return to consider a model of a finite number of stations in Section 5.4.

Exercises

Exercise 5.1 In this exercise we show that if we can accomplish some positive throughput, with longer messages we can accomplish a throughput that is arbitrarily close to 1.

Suppose that a message comprises a single "random access" packet and a random number K of data packets. A station that successfully transmits a random access packet is able to reserve K available slots for the rest of the message. (For example, the station might announce it will use the next K slots to transmit the data packets, and all the other stations avoid transmitting during those slots, as in Figure 5.4. Or the channel might be time divided, so that every kth slot is used by random access packets, and a successful random access packet reserves slots at times not divisible by k.) Indicate why, if a random access scheme achieves a stable throughput

Figure 5.4 A sequence of "random access" and "data" packets from two stations. White transmits one random access packet followed by the rest of the message; three random access packets collide; grey successfully transmits; white transmits again.

$\eta > 0$ for the random access packets, a stable throughput of

$$\frac{\mathbb{E}K}{\mathbb{E}K + \eta^{-1}}$$

should be expected in such a hybrid system.

Exercise 5.2 In this exercise we look briefly at a distributed model of random access, with a finite number of stations. We have seen a model of this in continuous time, in Example 2.24. We now describe a variant of this model in discrete time, to allow for delay in sensing a transmission. Recall that the interference graph has as vertices the stations, and stations i and j are joined by an edge if the two stations interfere with each other: write $i \sim j$ in this case. First suppose there is an infinite backlog of packets to transmit at each station, and suppose that station j transmits a packet in a time slot with probability f_j, independently over all time slots and all stations. Deduce that successful transmissions from station j form a Bernoulli sequence of rate

$$f_j \prod_{i \sim j} (1 - f_i).$$

Hence deduce that if, for each j, packet arrivals at station j occur at a rate less than this and station j transmits with probability f_j if it has packets queued, the system will be stable.

5.2 Estimating backlog

The problem with the ALOHA protocol is that when the backlog becomes large, the probability of successfully transmitting a packet becomes very small because packets almost invariably collide. Now, if the stations knew the backlog (they don't!), they could reduce their probability of retransmission when the backlog becomes large, thus reducing the number of collisions.

Let us now not distinguish between old and new packets, and simply let N_t be the number of packets that await to be transmitted. Suppose that each of them will be transmitted or not with the same probability f, and let us compute the optimal probability of retransmission for a given backlog size. The probability of exactly one transmission attempt given that the backlog is $N_t = n$ is then $nf(1 - f)^{n-1}$. If we maximize this with respect to f, we obtain

$$0 = \frac{\partial}{\partial f} = n(1 - f)^{n-1} - n(n-1)f(1 - f)^{n-2} \implies f = \frac{1}{n},$$

in which case the probability of a single retransmission attempt is given by

$$\left(1 - \frac{1}{n}\right)^{n-1} \to e^{-1} \qquad \text{as } n \to \infty.$$

Thus, if the stations knew the size of the backlog, we would expect them to be able to maintain a throughput of up to $1/e \approx 0.368$.

Now suppose that stations don't know the backlog size, but can observe the channel state. We assume that all stations are observing the channel from time 0, so they have access to the sequence $Z_1, Z_2, \ldots$. They will maintain a counter S_t (which we hope to use as a proxy for N_t), the same for all stations, even those with nothing to transmit. Once a station has a packet to transmit, it will attempt transmission with probability $1/S_t$.

Intuitively, the estimate S_t should go up if $Z_t = *$, because a collision means that stations are transmitting too aggressively. It should go down if $Z_t = 0$, because that means the stations are not being aggressive enough, and have wasted a perfectly good time slot for transmissions. If $Z_t = 1$, we might let S_t decrease (there was a successful transmission), or stay constant (the transmission probability is about right). So, for example, we might try

$$S_{t+1} = \begin{cases} \max(1, S_t - 1), & Z_t = 0 \text{ or } 1 \\ S_t + 1, & Z_t = * \end{cases} \quad \text{(naïve values)}. \tag{5.1}$$

(We don't want S_t to fall below 1 because we are using $1/S_t$ as the transmission probability.) More generally, we will let

$$S_{t+1} = \max(1, S_t + aI\{Z_t = 0\} + bI\{Z_t = 1\} + cI\{Z_t = *\}) \tag{5.2}$$

for some triplet (a, b, c), where we expect $a < 0$ and $c > 0$.

Our hope is that, if S_t can track N_t, we may be able to achieve a positive throughput. Note that N_t alone and S_t alone are not Markov chains, but the pair (S_t, N_t) is a Markov chain. (Why?) We next compute the drift of this Markov chain.

Computing the probability of 0, 1, or greater than 1 transmission attempts during a time slot, we have (check these!)

$$\begin{aligned} \mathbb{E}[S_{t+1} - S_t \mid S_t = s, N_t = n] &= a\left(1 - \frac{1}{s}\right)^n + b\left(\frac{n}{s}\right)\left(1 - \frac{1}{s}\right)^{n-1} \\ &\quad + c\left(1 - \left(1 - \frac{1}{s}\right)^n - \left(\frac{n}{s}\right)\left(1 - \frac{1}{s}\right)^{n-1}\right) \end{aligned}$$

and

$$\mathbb{E}[N_{t+1} - N_t \mid S_t = s, N_t = n] = \nu - \frac{n}{s}\left(1 - \frac{1}{s}\right)^{n-1}.$$

Consider now a deterministic approximation to our system by a pair of coupled differential equations; i.e. trajectories $(s(t), n(t))$ with

$$\frac{ds}{dt} = \mathbb{E}[S_{t+1} - S_t \mid S_t = s, N_t = n], \quad \frac{dn}{dt} = \mathbb{E}[N_{t+1} - N_t \mid S_t = s, N_t = n].$$

We shall investigate the drifts as a function of $\kappa(t) = n(t)/s(t)$, for large values of n, s when, in the limit,

$$\frac{ds}{dt} = (a-c)e^{-\kappa} + (b-c)\kappa e^{-\kappa} + c, \qquad \frac{dn}{dt} = \nu - \kappa e^{-\kappa}. \tag{5.3}$$

We aim to show that trajectories of solutions to these differential equations converge towards the origin, as in Figure 5.5.

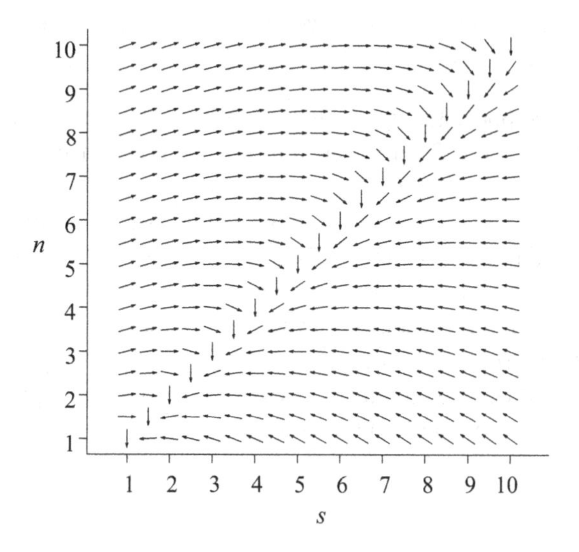

Figure 5.5 Vector field of (s, n) when $\nu = 1/3$ and $(a, b, c) = (1 - e/2, 1 - e/2, 1)$.

Suppose we choose (a, b, c) so that

$$(a-c)e^{-\kappa} + (b-c)\kappa e^{-\kappa} + c \begin{cases} < 0, & \kappa < 1, \\ > 0, & \kappa > 1. \end{cases} \tag{5.4}$$

Then $ds/dt <$ or > 0 according as $\kappa(t) <$ or > 1, and trajectories cross the diagonal $\kappa = 1$ vertically, and in a downwards direction provided $\nu < 1/e$. This looks promising, since once the trajectory is below $\kappa = 1$ we have $ds/dt < 0$. But we need to know more when $\kappa(t) > 1$.

Suppose we insist that (a, b, c) also satisfy

$$\nu - \kappa e^{-\kappa} < \kappa\left((a-c)e^{-\kappa} + (b-c)\kappa e^{-\kappa} + c\right), \qquad \text{for } \kappa > 1. \tag{5.5}$$

Then (Exercise 5.3) the derivative $d\kappa(t)/dt < 0$ for $\kappa(t) > 1$, so that in the region $\kappa > 1$ trajectories approach and cross the diagonal $\kappa = 1$.

For either of the choices

$$(a, b, c) = (2 - e, 0, 1) \quad \text{or} \quad (1 - e/2, 1 - e/2, 1) \tag{5.6}$$

(or many other choices), and any value of $\nu < e^{-1}$, both of these conditions are satisfied (Exercise 5.3), and we can deduce that all trajectories approach the origin. In Figure 5.5, we plot the vector field when $(a, b, c) = (1 - e/2, 1 - e/2, 1)$ and $\nu = 1/3$.

Remark 5.7 Suppose the triplet (a, b, c) satisfies the conditions (5.4) and (5.5), and that $\nu < 1/e$. Then, along a trajectory, $\kappa(t)$ converges to a value $\kappa_\nu < 1$, and $\kappa_\nu \uparrow 1$ as $\nu \uparrow 1/e$.

The triplet $(a, b, c) = (-1, -1, 1)$ suggested in (5.1) does not manage to satisfy the conditions (5.4) and (5.5), and its maximal throughput is less than $1/e$ (but still positive). In Figure 5.6, we show the vector field of the equation for $\nu = 1/3$ (now unstable) and $\nu = 1/6$ (stable).

Note that the triplet $(a, b, c) = (1 - e/2, 1 - e/2, 1)$ only requires the stations to know whether a collision occurred – it does not require a station to distinguish between channel states 0 and 1.

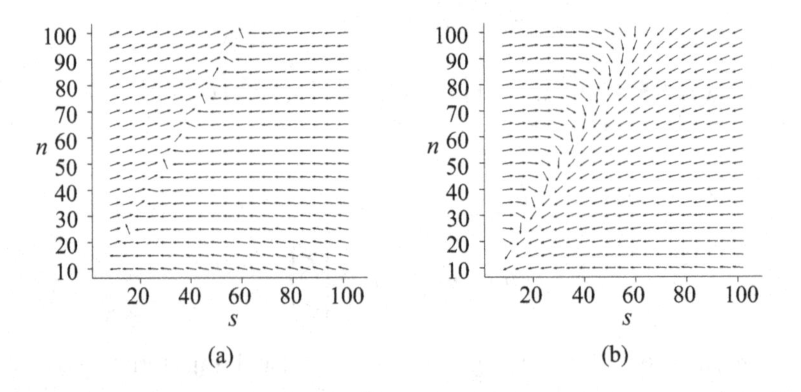

Figure 5.6 Vector field of (s, n) with the "naïve" parameters $(a, b, c) = (-1, -1, 1)$. (a) $\nu = 1/3$ (now unstable). (b) $\nu = 1/6$ (stable).

Just as in Remark 5.2 drift analysis did not conclusively prove transience, here we have not conclusively proven positive recurrence yet. Negative drift does mean that the chain will not be transient (Meyn and Tweedie

(1993) Theorem 8.4.3), but the expected return time may be infinite, as the following argument easily shows. Consider a Markov chain that jumps from 0 to $2^n - 1$, for $n \geq 1$, with probability 2^{-n}, and for $n > 1$ transitions deterministically $n \to n - 1$. Then, starting from 0, the return time will be 2^n with probability 2^{-n}, and so the expected return time will be infinite.

We will prove positive recurrence of this system later, in Exercise D.3. Rather than finding an explicit equilibrium distribution (which would be quite hard in this case), we will use the *Foster–Lyapunov criteria*, a stricter version of the drift conditions.

Exercises

Exercise 5.3 Show that condition (5.5) is sufficient to ensure that $\kappa(t)$ is decreasing when $\kappa(t) > 1$.

Show that either of the choices (5.6) satisfy both conditions (5.4) and (5.5) provided $\nu < 1/e$.

Exercise 5.4 Show that if $c \leq 0$ then the conclusion of Proposition 5.3 holds – the channel jams after a finite time – for the scheme (5.1).

Exercise 5.5 Repeat the argument of this section with the "naïve" values $a = -1$, $b = -1$, $c = 1$ of (5.1). Check that the drift of S is negative if $N_t < \kappa S_t$ and positive if $N_t > \kappa S_t$, for $\kappa \approx 1.68$. What throughput should this system allow?

5.3 Acknowledgement-based schemes

Often, it is not possible for stations to observe the channel. For example, a device using the Internet typically learns that a sent packet has been successfully received when it receives an *acknowledgement packet*, or *ACK*. We will now consider what happens when the only information a station has concerning other stations or the use of the channel is the history of its own transmission attempts.

Let us model the situation as follows. Packets arrive from time 0 onwards in a Poisson stream of rate $\nu > 0$. A packet arriving in slot t will attempt transmission in slots $t + x_0, t + x_1, t + x_2, \ldots$ until the packet is successfully transmitted, where $1 = x_0 < x_1 < x_2 < \ldots$, and $X = (x_1, x_2, \ldots)$ is a random (increasing) sequence. Assume the choice of the sequence X is independent from packet to packet and of the arrival process. (We could imagine the packet arrives equipped with the set of times at which it will attempt retransmission.) Write $h(x)$ for the probability that $x \in X$, so that $h(1) = 1$.

Example 5.8 (ALOHA) For the ALOHA protocol of Section 5.1, $h(x) = f$ for all $x > 1$.

Example 5.9 (Ethernet) The *Ethernet* protocol is defined as follows. After r unsuccessful attempts, a packet is retransmitted after a period of time chosen uniformly from the set $\{1, 2, 3, \ldots, 2^r\}$. (This is called *binary exponential backoff*.)

Here, there isn't an easy expression for $h(x)$, but we can enumerate the first few values. As always, $h(1) = 1$; $h(2) = 1/2$; $h(3) = 5/8$; $h(4) = 17/64$; $h(5) = 305/1024$; and so on. (Check these.)

We are interested in the question of whether, like ALOHA, the channel will eventually jam forever, or if we can guarantee an infinite number of successful transmissions.

For the purpose of argument, suppose that the channel is externally jammed from time 0 onwards, so that no packets are successfully transmitted, and hence *all* potential packet retransmissions of the arriving packets occur. Let Y_t be the number of newly arrived packets that arrive during slot $t-1$ and are first transmitted in slot t: then $Y_1, Y_2, \ldots$ are independent Poisson random variables with mean ν.

Since the packets' retransmission sequences X are independent, the number of transmissions occurring in slot t will be Poisson with mean $(h(1) + \ldots + h(t))\nu$. (The number of packets arriving in slot 0 and transmitting in slot t is Poisson with mean $\nu h(t)$, the number of packets arriving in slot 1 and transmitting in slot t is Poisson with mean $\nu h(t-1)$, and so on.)

The probability of fewer than two attempts in any given time slot t is

$$P_t = \mathbb{P}\Big(\text{Poisson}((h(1) + \ldots + h(t))\nu) = 0 \text{ or } 1\Big) = \left(1 + \nu \sum_{r=1}^{t} h(r)\right) e^{-\nu \sum_{r=1}^{t} h(r)}.$$

Let $\overline{\Phi}$ be the set comprising those slots in which less than two attempts are made in this externally jammed channel. Then the expected number of such slots is

$$H(\nu) \equiv \mathbb{E}|\overline{\Phi}| = \sum_{t=1}^{\infty} P_t.$$

(The numbers of attempts in different slots are not independent, but the linearity of expectation is all we've used here.) The probabilities P_t, and hence $H(\nu)$, are decreasing in ν. Let

$$\nu_c = \inf\{\nu : H(\nu) < \infty\}.$$

Thus, if $\nu > \nu_c$ then $\mathbb{E}|\overline{\Phi}| < \infty$, and hence $\overline{\Phi}$ is finite with probability 1.

We will first show that ν_c is the critical rate: i.e. for arrival rates below ν_c, with probability 1 infinitely many packets are successfully transmitted; for arrival rates above ν_c, only finitely many packets are. We will then compute the value of ν_c for some schemes, including ALOHA and Ethernet.

To proceed with the argument, now remove the external jamming, and let Φ be the set comprising those slots in which fewer than two attempts are made. Then $\overline{\Phi} \subset \Phi$, since any element of $\overline{\Phi}$ is necessarily an element of Φ. We will now show that, if $\nu > \nu_c$, then with positive probability the containment is equality.

Consider an arrival rate $\nu > \nu_c$. Subject the system to an additional, independent, Poisson arrival stream of rate ϵ. There is a positive probability that the additional arrivals jam every slot in the set $\overline{\Phi}$, since $\overline{\Phi}$ is finite with probability 1. Thus there is positive probability, say $p > 0$, that in *every* slot two or more transmissions are attempted, even without any external jamming: you should convince yourself that in this case $\Phi = \overline{\Phi} = \emptyset$.

Suppose that did not occur, and consider the situation just after the first slot, say slot τ_1, in which either 0 or 1 transmissions are attempted. It looks just like time 0, except we likely have a backlog of packets already present. These packets only create more collisions; so the probability that the channel will never unjam from time τ_1 onwards is at least p.

Similarly, following each time slot in which either 0 or 1 transmissions are attempted, the probability that the system jams forever from this time onwards is $\geq p$. Therefore, the number of successful packet transmissions is bounded above by a geometric random variable with parameter $1 - p$. Since ν and ϵ are arbitrary subject to $\nu > \nu_c$, $\epsilon > 0$, it follows that for a system with arrival rate $\nu > \nu_c$ the expected number of successful packet transmissions is finite. In particular, the number of successful packet transmissions is finite with probability 1.

Remark 5.10 This style of argument is known as *coupling*: in order to show that the jamming probability from time τ_1 onwards is at least p, we consider realizations of "system starting empty" and "system starting with a backlog" where pathwise the backlog can only create more collisions.

Next we show that if $\nu_c > 0$ and the arrival rate satisfies $\nu < \nu_c$, then the set Φ is infinite with probability 1. Suppose otherwise: then there is positive probability that the set Φ is finite. Now subject the system to an additional, independent, Poisson arrival stream of rate ϵ, where $\nu + \epsilon < \nu_c$. Then there is a positive probability, say $p > 0$, that in *every* slot two or more transmissions are attempted. Indeed, the earlier coupling argument shows that the number of such slots is bounded by a geometric random variable with parameter $1 - p$. But this then implies that, for the system with arrival rate $\nu+\epsilon$, $\mathbb{E}\,|\Phi| < \infty$ and hence $\mathbb{E}\left|\overline{\Phi}\right| < \infty$. But this contradicts the definition of ν_c. Hence we deduce that the set Φ is infinite with probability 1 for any arrival rate $\nu < \nu_c$.

We have shown that the channel will unjam infinitely many times for $\nu < \nu_c$, but does this imply that there are infinitely many successful transmissions? To address this question, let Ψ_i be the set comprising those slots where i *re*transmission attempts are made, for $i = 0, 1$. Then $\Phi \subset \Psi_0 \cup \Psi_1$ (the inclusion may be strict, since the definition of Ψ_0, Ψ_1 does not involve first transmission attempts). Now,

$$|\Phi| = \sum_{t=1}^{\infty} (I[t \in \Psi_1, Y_t = 0] + I[t \in \Psi_0, Y_t = 1] + I[t \in \Psi_0, Y_t = 0]).$$

But Y_t is independent of $I[t \in \Psi_0]$, and so, with probability 1,

$$\sum_{t=1}^{\infty} (I[t \in \Psi_0, Y_t = 1] + I[t \in \Psi_0, Y_t = 0]) = \infty$$

$$\implies \sum_{t=1}^{\infty} I[t \in \Psi_0, Y_t = 1] = \infty.$$

Since $|\Phi| = \infty$ with probability 1 we deduce that

$$\sum_{t=1}^{\infty} (I[t \in \Psi_1, Y_t = 0] + I[t \in \Psi_0, Y_t = 1]) = \infty$$

with probability 1.

We have thus shown the following result.

Theorem 5.11 *If $\nu \in (0, \nu_c)$ then with probability 1 an infinite number of packets are successfully transmitted.*

If $\nu > \nu_c$ then with probability 1 only a finite number of packets are successfully transmitted; further, the expected number of packets successfully transmitted is finite.

Thus, the question we would like to answer is: which back-off schemes (equivalently, choices of the function $h(x)$) have $\nu_c > 0$?

Example 5.12 (ALOHA) Recall that, for the ALOHA scheme, $h(x) = f$ for all $x \geq 2$, and so $P_t \sim \nu f t e^{-\nu f t}$. Hence $H(\nu) < \infty$ for any $\nu > 0$. Thus $\nu_c = 0$; as we have already seen, with any positive arrival rate the system will eventually jam. More generally, $\nu_c = 0$ whenever

$$(\log t)^{-1} \sum_{x=1}^{t} h(x) \to \infty \qquad \text{as} \quad t \to \infty, \tag{5.7}$$

since this implies $\sum_{t=1}^{\infty} P_t < \infty$ for any $\nu > 0$ (Exercise 5.8). In this sense, the expected number of successful transmissions is finite for any acknowledgement-based scheme with slower than exponential backoff.

Example 5.13 (Ethernet) For the Ethernet protocol described above, as $t \to \infty$,

$$\sum_{r=1}^{t} h(r) \sim \log_2 t.$$

(The expected time for a packet to make k retransmission attempts is roughly 2^k.) Therefore,

$$e^{-\nu \sum_{r=1}^{t} h(r)} = t^{-\nu/\log 2 + o(1)},$$

hence

$$P_t = \nu t^{-\nu/\log 2 + o(1)} \log_2 t,$$

and $\sum_{t=1}^{\infty} P_t < \infty$ or $= \infty$ according as $\nu >$ or $< \log 2 \approx 0.693$. That is, $\nu_c = \log 2$.

Remark 5.14 Note that this does not mean that the Markov chain describing the backlog of packets is positive recurrent. Indeed, suppose we had a positive recurrent system, with π_0 the equilibrium probability of 0 retransmissions, and π_1 the equilibrium probability of 1 retransmission in any given time slot. Then the expected number of (new and old) packet transmissions in a slot is

$$\pi_1 e^{-\nu} + \pi_0 \nu e^{-\nu},$$

while the expected number of arrivals is ν. Now, $\pi_0 + \pi_1 \leq 1$, and clearly $\nu \leq 1$, so $\pi_1 e^{-\nu} + \pi_0 \nu e^{-\nu} \leq e^{-\nu}$, and we obtain

$$\nu \leq e^{-\nu} \implies \nu \leq 0.567.$$

Thus, for $0.567 < \nu < 0.693$, an acknowledgement-based scheme is most definitely not positive recurrent, even though Ethernet will have an infinite number of packets successfully transmitted in this range.

In fact, (a geometric version of) the Ethernet scheme will not be positive recurrent for any positive arrival rate.

Theorem 5.15 (Aldous (1987)) *Suppose the backoff after r unsuccessful attempts is distributed geometrically with mean 2^r. (This will give the same asymptotic behaviour of $h(t)$, but a nicer Markov chain representation.) Then the Markov chain describing the backlog of untransmitted packets is transient for all $\nu > 0$. Further, if $N(t)$ is the number of packets successfully transmitted by time t, then with probability* 1, $N(t)/t \to 0$ *as* $t \to \infty$.

It is not known whether there exists any acknowledgement-based scheme and arrival rate $\nu > 0$ such that the scheme is positive recurrent. Theorem 5.15 suggests not, and that some form of channel sensing or feedback, as in Section 5.2, is necessary for positive recurrence. (In practice, acknowledgement-based schemes generally discard packets after a finite number of attempts.)

Remark 5.16 The models of this chapter have assumed that the time axis can be slotted. In view of the very limited amount of information available to a station under an acknowledgement-based scheme, it is worth noting what happens without this assumption. Suppose that a packet that arrives at time $t \in \mathbb{R}_+$ is first transmitted in the interval $(t, t+1)$; the transmission is unsuccessful if any other station transmits for any part of the interval, and otherwise the transmission is successful. Suppose that, after r unsuccessful attempts, the packet is retransmitted after a period of time chosen uniformly from the real interval $(1, \lfloor b^r \rfloor)$. Then it is known that the probability that infinitely many packets are transmitted successfully is 1 or 0 according as $\nu \leq \nu_c$ or $\nu > \nu_c$, where $\nu_c = \frac{1}{2} \log b$ (Kelly and MacPhee (1987)). Comparing this with Exercise 5.7, we see that moving from a slotted to a continuous time model halves the critical value ν_c.

Exercises

Exercise 5.6 Theorem 5.11 left open the case $\nu = \nu_c$. Show that this case can be decided according to whether $H(\nu_c)$ is finite or not.
[*Hint:* Is the additional Poisson arrival stream of rate ϵ really needed in the proof of the theorem?]

Exercise 5.7 Consider the following generalization of the Ethernet protocol. Suppose that, after r unsuccessful attempts, a packet is retransmitted after a period of time chosen uniformly from $\{1, 2, 3, \ldots, \lfloor b^r \rfloor\}$. Thus $b = 2$ corresponds to binary exponential backoff. Show that $\nu_c = \log b$.

Exercise 5.8 Check the claim that $\nu_c = 0$ if condition (5.7) is satisfied. Show that if

$$\sum_{x=1}^{\infty} h(x) < \infty$$

a packet is transmitted at most a finite number of times (equivalently, it is discarded after a finite number of attempts), and that $\nu_c = \infty$.
[*Hint:* First Borel–Cantelli lemma.]

Suppose that a packet is retransmitted a time x after its first transmission with probability $h(x)$, independently for each x until it is successful. Show that if

$$h(x) = \frac{1}{x \log x}, \qquad x \geq 2,$$

then no packets are discarded and $\nu_c = \infty$.

Exercise 5.9 Any shared information between stations may aid coordination. For example, suppose that in an acknowledgement-based scheme a station can distinguish even from odd slots, according to a universal clock, after its first transmission. Show that, if a station transmits for the first time as soon as it can, but confines retransmission attempts to even slots, then the number of packets successfully transmitted grows at least at rate $\nu e^{-\nu}/2$.

5.4 Distributed random access

In this section we consider a finite number of stations attempting to share a channel using the ALOHA protocol, where not all stations interfere with each other. We begin by supposing stations form the vertex set of an interference graph, with an edge between two stations if they interfere with each other. This would be a natural representation of a wireless network distributed over a geographical area.

We have seen already in Example 2.24 a simple model operating in continuous time. Recalling that model, let $\mathcal{S} \subset \{0, 1\}^R$ be the set of vectors $n = (n_r, r \in \mathcal{R})$ with the property that if i and j have an edge between them then $n_i \cdot n_j = 0$. Let n be a Markov process with state space $\mathcal{S}$ and transition rates

$$\begin{aligned} q(n, n - e_r) &= 1 && \text{if} \quad n_r = 1, \\ q(n, n + e_r) &= \exp(\theta_r) && \text{if} \quad n + e_r \in \mathcal{S}, \end{aligned}$$

for $r \in \mathcal{R}$, where $e_r \in \mathcal{S}$ is a unit vector with a 1 as its rth component and

0s elsewhere. We interpret $n_r = 1$ as indicating that station r is transmitting, and we assume that other stations within interference range of station r (i.e. stations that share an edge with station r) sense this and do not transmit. Station r transmits for a time that is exponentially distributed with unit mean. If $n_r = 0$ and no station is already transmitting within interference range of station r, then station r starts transmitting at rate $\exp(\theta_r)$. We ignore for the moment any delays in sensing, and thus any collisions.

The equilibrium distribution for $n = (n_r, r \in \mathcal{R})$ can, from Example 2.24 or Section 3.3, be written in the form

$$\pi(n) = \frac{\exp(n \cdot \theta)}{\sum_{m \in \mathcal{S}} \exp(m \cdot \theta)}, \qquad n \in \mathcal{S}, \tag{5.8}$$

where $\theta = (\theta_r, r \in \mathcal{R})$. The equilibrium probability that station r is transmitting is then

$$\mathbb{E}_\theta[n_r] = \sum_{n \in \mathcal{S}} n_r \pi(n) = \frac{\sum_{n \in \mathcal{S}} n_r \exp(n \cdot \theta)}{\sum_{m \in \mathcal{S}} \exp(m \cdot \theta)}. \tag{5.9}$$

If we imagine that packets arrive at station r at a mean rate λ_r and are served when station r is transmitting, we should expect the system to be stable provided $\lambda_r < \mathbb{E}_\theta[n_r]$ for $r \in \mathcal{R}$. By varying the parameters $\theta = (\theta_r, r \in \mathcal{R})$, what range of values for $\lambda = (\lambda_r, r \in \mathcal{R})$ does this allow?

Let

$$\Lambda = \left\{ \lambda \geq 0 : \exists p(n) \geq 0, \sum_{n \in \mathcal{S}} p(n) = 1 \right.$$
$$\left. \text{such that } \sum_{n \in \mathcal{S}} p(n) n_r \geq \lambda_r \text{ for } r \in \mathcal{R} \right\}.$$

We now show we can allow any vector in the interior of Λ.

Theorem 5.17 (Jiang and Walrand (2010)) *Provided λ lies in the interior of the region Λ there exists a vector θ such that*

$$\mathbb{E}_\theta[n_r] > \lambda_r, \qquad r \in \mathcal{R},$$

where $\mathbb{E}_\theta[n_r]$ is given by equation (5.9).

Proof Consider the optimization problem

$$\begin{aligned} &\text{minimize} && \sum_{n\in\mathcal{S}} p(n)\log p(n) && && (5.10)\\ &\text{subject to} && \sum_{n\in\mathcal{S}} p(n)n_r \geq \lambda_r, && r\in\mathcal{R}, \\ &\text{and} && \sum_{n\in\mathcal{S}} p(n) = 1 \\ &\text{over} && p(n)\geq 0, \quad n\in\mathcal{S}. \end{aligned}$$

Since λ lies in the interior of Λ, there exists a feasible solution to this problem, and indeed there exists a feasible solution for any sufficiently small perturbation of the right-hand sides of the constraints. Further, the optimum is attained, since the feasible region is compact. The objective function is convex and the constraints are linear, and so the strong Lagrangian principle holds (Appendix C).

The Lagrangian for the problem is

$$L(p,z;\theta,\kappa)$$
$$= \sum_{n\in\mathcal{S}} p(n)\log p(n) + \sum_{r\in\mathcal{R}} \theta_r\left(\lambda_r + z_r - \sum_{n\in\mathcal{S}} p(n)n_r\right) + \kappa\left(1 - \sum_{n\in\mathcal{S}} p(n)\right),$$

where $z_r, r\in\mathcal{R}$, are slack variables and $\kappa, \theta_r, r\in\mathcal{R}$, are Lagrange multipliers for the constraints. We now attempt to minimize L over $p(n)\geq 0$ and $z_r \geq 0$. To obtain a finite minimum over $z_r \geq 0$, we require $\theta_r \geq 0$; and given this we have at the optimum $\theta_r \cdot z_r = 0$. Further, differentiating with respect to $p(n)$ gives

$$\frac{\partial L}{\partial p(n)} = 1 + \log p(n) - \sum_{r\in\mathcal{R}} \theta_r n_r - \kappa.$$

By the strong Lagrangian principle, we know there exist Lagrange multipliers $\kappa, \theta_r, r\in\mathcal{R}$, such that the Lagrangian is maximized at p, z that are feasible, and p, z are then optimal, for the original problem.

At a minimum over $p(n)$,

$$p(n) = \exp\left(\kappa - 1 + \sum_{r\in\mathcal{R}} \theta_r n_r\right).$$

Choose κ so that $(p(n), n\in\mathcal{S})$ sum to 1: then

$$p(n) = \frac{\exp(\sum_{r\in\mathcal{R}} \theta_r n_r)}{\sum_{m\in\mathcal{S}} \exp(\sum_{r\in\mathcal{R}} \theta_r m_r)},$$

precisely of the form (5.8). Since $z_r \geq 0$, we have found a vector θ such that

$$\sum_{n \in \mathcal{S}} p(n) n_r \geq \lambda_r,$$

with equality if $\theta_r > 0$. To obtain the strict inequality of the theorem, note that, since λ lies in the interior of the region Λ, we can use a perturbation of λ in the above construction.

□

Remark 5.18 If we had used the transition rates

$$q(n, n - e_r) = \exp(-\theta_r) \quad \text{if} \quad n_r = 1,$$
$$q(n, n + e_r) = 1 \quad \text{if} \quad n + e_r \in \mathcal{S},$$

for $r \in \mathcal{R}$, the equilibrium distribution (5.8) would be unaltered. Under the first variant, stations transmit for a period with unit mean, whatever the the value of θ, while under this second variant a station attempts to start transmission at unit rate whatever the value of θ. A disadvantage of the first variant is that if components of θ are large then collisions (caused by a delay in sensing that a station within interference range has also started transmitting) will become more frequent. The second variant avoids this, and so, although transmissions may occasionally collide, there is no tendency for this to increase as loads increase. But a disadvantage of the second variant is that if components of θ are large then stations may hog the system, blocking other stations, for long periods of time.

If we could scale up all the transition rates by the same factor, this would reduce the hogging periods by the same factor; but of course speed-of-light delays limit this.

Remark 5.19 The objective function (5.10) is (minus) the *entropy* of the probability distribution $(p(n), n \in \mathcal{S})$, and so the optimization problem finds the distribution of maximum entropy subject to the constraints. This functional form of objective produces the product-form (5.8) for p at an optimum. We observed something formally similar in Theorem 3.10, where the solution to the PRIMAL problem (3.4) had a product-form. In each case there are dual variables for each constraint, although these constraints are of very different forms. As in the earlier problem, there are many variations on the objective function (5.10) that will leave the optimum of product-form, and we consider one in Exercise 5.11.

Example 5.20 (Optical wavelength-routed networks) So far in this section we have defined the state space $\mathcal{S}$ in terms of an interference graph, but all that is needed is that $\mathcal{S}$ be a subset of $\{0, 1\}^R$, with the following hierarchical property: $n + e_r \in \mathcal{S} \implies n \in \mathcal{S}$. As an example we describe a simple model of an optical wavelength-routed network. Such networks are able to provide very substantial capacity in the core of a communication network.

In optical networks each fibre is partitioned into a number of wavelengths, each capable of transmitting data. The fibres are connected by switches (called optical cross-connects): the fibres are the edges and the switches the nodes of the physical topology. A connection across the network must be routed over the physical topology and must also be assigned a wavelength. The combination of physical route and wavelength is known as a *lightpath*. We consider an all-optical network, where the same wavelength is used at each edge along the path (if the wavelength can be converted at intermediate nodes, the model resembles a loss network, as in Chapter 3).

Lightpaths carry data between endpoints of the physical topology, providing capacity between pairs of endpoints. But how should lightpaths be provided, so that the capacity between endpoints is aligned with demand?

Suppose that on each fibre there are the same set of (scalar) C wavelengths, and let r label a lightpath. We suppose r identifies a source–sink pair $s(r)$, a set of fibres and a wavelength, and we let $\mathcal{R}$ be the set of lightpaths. It will be convenient to write $r \in s$ if source–sink pair s is identified by lightpath r. Note that we do *not* assume that all lightpaths $r \in s$ identify the same set of fibres: a source–sink pair may be served by multiple paths through the physical topology, as well as by multiple wavelengths on a given path.

Let $n_r = 0$ or 1 indicate that lightpath r is inactive or active, respectively, and let $n = (n_r, r \in \mathcal{R})$. Then a state is feasible if active lightpaths using the same wavelength have no fibres in common: let $\mathcal{S}(C) \subset \{0, 1\}^R$ be the set of feasible states. With a mild risk of confusion, we let $\mathcal{S}$ be the set of source–sink pairs.

Suppose that each source–sink pair s maintains a parameter θ_s, and suppose that n is a Markov process with state space $\mathcal{S}(C)$ and transition rates

$$\begin{aligned} q(n, n - e_r) &= 1 && \text{if} \quad n_r = 1, \\ q(n, n + e_r) &= \exp(\theta_{s(r)}) && \text{if} \quad n + e_r \in \mathcal{S}(C), \end{aligned}$$

for $r \in \mathcal{R}$, where we recall that $e_r \in \mathcal{S}(C)$ is a unit vector with a 1 as its rth component and 0s elsewhere.

We recognize this model as just another example of a truncated reversible process, and its stationary distribution is given by expression (5.8), where $\theta_r = \theta_{s(r)}$. But what range of loads $\lambda = (\lambda_s, s \in \mathcal{S})$ does this allow?

Let

$$\Lambda = \left\{ \lambda \geq 0 : \exists p(n) \geq 0, \sum_{n \in \mathcal{S}(C)} p(n) = 1 \right.$$
$$\left. \text{such that} \sum_{r \in \mathcal{R}} \sum_{n \in \mathcal{S}(C)} p(n) n_r \geq \lambda_s \text{ for } s \in \mathcal{S} \right\}.$$

In Exercise 5.14 it is shown that $\theta = (\theta_s, s \in \mathcal{S})$ can be chosen to support any vector in the interior of Λ.

Remark 5.21 If we imagine time is slotted, we can multiply wavelengths in the above model by time as well as frequency division. This might help if we want to increase the ratio of raw wavelengths to source–sink pairs.

We have used the properties of a wavelength-routed network to motivate a particular model of routing, and such models may arise in a variety of other contexts. We should note that wide-area optical wavelength-routed networks may provide the basic link capacities in a loss network, and in this role we would expect the allocation of lightpaths to source–sink pairs to change rather slowly, perhaps over a daily cycle. But the above discussion does suggest that simple decentralized algorithms driven by random local choices have the potential to be remarkably effective.

Exercises

Exercise 5.10 By substituting the values of p and z optimizing the Lagrangian, show that the dual to problem (5.10) is

$$\text{maximize} \quad \mathcal{V}(\theta) = \sum_{r \in \mathcal{R}} \lambda_r \theta_r - \log\left(\sum_{n \in \mathcal{S}} \exp\left(\sum_{r \in \mathcal{R}} \theta_r n_r \right) \right)$$
$$\text{over} \quad \theta_r \geq 0, \quad r \in \mathcal{R}.$$

Show that

$$\frac{\partial \mathcal{V}}{\partial \theta_r} = \lambda_r - \mathbb{E}_\theta[n_r],$$

where the expectation is calculated under the distribution (5.8). (We shall use this result later, in Exercise 7.20.)

Exercise 5.11 If we imagine station r as a server for the packets that arrive there, then a very crude model for the mean number of packets queued at station r is $\lambda_r/(\mathbb{E}[n_r] - \lambda_r)$: this is the approximation we used in Example 2.22, leading to Kleinrock's square root assignment. Consider the problem

$$\begin{aligned}
&\text{minimize} && \sum_{n\in\mathcal{S}} p(n)\log p(n) + D\sum_{r\in\mathcal{R}} \frac{\lambda_r}{z_r} \\
&\text{subject to} && \sum_{n\in\mathcal{S}} p(n)n_r = \lambda_r + z_r, \qquad r\in\mathcal{R}, \\
&\text{and} && \sum_{n\in\mathcal{S}} p(n) = 1 \\
&\text{over} && p(n)\geq 0, \quad n\in\mathcal{S}; \qquad z_r\geq 0, \quad r\in\mathcal{R}.
\end{aligned}$$

We have added to the objective function our crude representation of the mean number of packets at stations, weighted by a positive constant D. Show that, provided λ lies in the interior of the region Λ, the optimum is again of the form (5.8).

Show that the dual to the optimization problem is

$$\begin{aligned}
&\text{maximize} && \mathcal{V}(\theta) = \sum_{r\in\mathcal{R}}\left(\lambda_r\theta_r + 2(D\lambda_r\theta_r)^{1/2}\right) - \log\left(\sum_{n\in\mathcal{S}}\exp\left(\sum_{r\in\mathcal{R}}\theta_r n_r\right)\right) \\
&\text{over} && \theta_r\geq 0, \qquad r\in\mathcal{R},
\end{aligned}$$

and that

$$\frac{\partial\mathcal{V}}{\partial\theta_r} = \lambda_r + \left(\frac{D\lambda_r}{\theta_r}\right)^{1/2} - \mathbb{E}_\theta[n_r].$$

Exercise 5.12 Consider the choice of the constant D in the model of Exercise 5.11. Write $\theta_r = \theta_r(D), r\in\mathcal{R}$, to emphasize the dependence of the Lagrange multipliers on D. From the form of the optimizing variables z, show that the mean number of packets at station r is

$$\left(\frac{\theta_r(D)\lambda_r}{D}\right)^{1/2}.$$

Deduce that, as $D\to\infty$, the form (5.8) concentrates probability mass on vectors n that each identify a maximal independent vertex set (i.e. an independent vertex set that is not a strict subset of any other independent vertex set).

Exercise 5.13 Show that it is possible to view the state space $\mathcal{S}(C)$ of Example 5.20 as derived from an interference graph, to be described.

Exercise 5.14 Establish the claim in Example 5.20.
[*Hint:* Consider the optimization problem (5.10), but with the first constraint replaced by

$$\sum_{r \in \mathcal{R}} \sum_{n \in \mathcal{S}(C)} p(n) n_r \geq \lambda_s \text{ for } s \in \mathcal{S}. \qquad]$$

5.5 Further reading

Our treatment, and use in Section 5.2, of the Foster–Lyapunov criteria follows Hajek (2006), which is an excellent text on several of the topics covered in this book. Goldberg *et al.* (2004) provide a review of early work on acknowledgement-based schemes; Jiang and Walrand (2012) and Shah and Shin (2012) discuss recent work on distributed random access.

The problem of decentralized communication becomes harder in a network setting, particularly if packets need to be transmitted across multiple links, or "hops". Fixed point approximations for networks with random access policies (in the spirit of the Erlang fixed point approximation) are discussed in Marbach *et al.* (2011). A popular alternative to distributed random access in the multihop setting is to use gossiping algorithms to coordinate the communications; Modiano *et al.* (2006) discuss this approach.

6

Effective bandwidth

Until now, our model of communication has assumed that an individual connection requests some amount of bandwidth (measured, for example, in circuits) and receives all or none of it. We then used this model to develop Erlang's formula and our analysis of loss networks. However, it is possible that the bandwidth profile of a connection is different, and in particular it might fluctuate. Thus, the peak rate required may be greater than the mean rate achieved over the holding period of the connection, as illustrated in Figure 6.1.

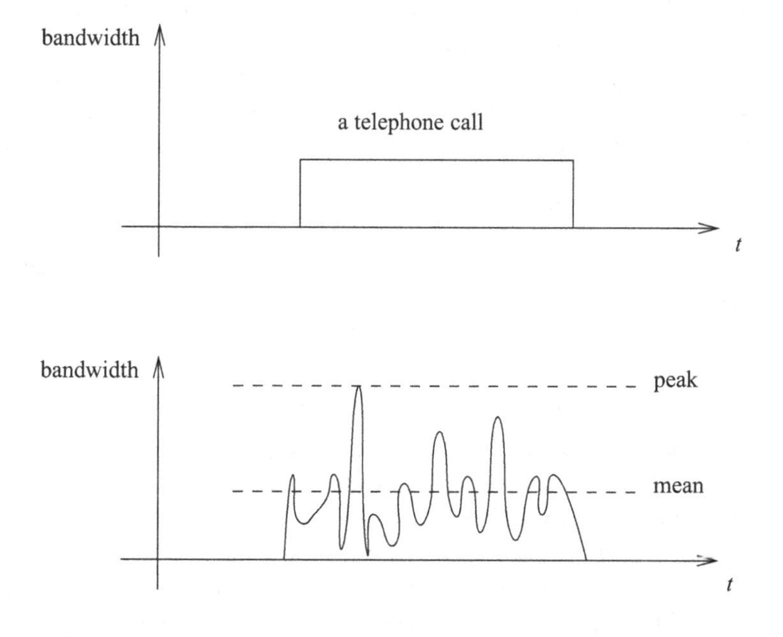

Figure 6.1 Possible bandwidth profiles; mean and peak rates.

Our goal in this chapter will be to understand how we might perform admission control for such a system, so as to achieve an acceptable balance between utilizing the capacity of the resources and the probability that the total required rate will exceed the capacity of a resource causing packets to be lost.

There are two extreme strategies that we can identify immediately, each of which reduces the problem to our earlier model of a loss network. At one extreme, we could assume that each connection has its peak rate reserved for it at each resource it passes through. This will mean that packets do not get lost, but may require a lot of excess capacity at resources. At the other extreme, we could reserve for each connection its mean rate and allow packets to queue for resources if there are instantaneous overloads. With intelligent management of queued packets, it may be possible to stabilize the resulting queueing network, but at the cost of large delays in packets being delivered. Our goal is to find a strategy between these two extremes.

Exercise

Exercise 6.1 Consider the following (over-)simplified model, based on Example 2.22. Suppose that a connection on route r produces a Poisson stream of packets at rate λ_r, and that a resource is well described as an $M/M/1$ queue with arrival rate $\sum_{r:j\in r} n_r\lambda_r$ and service rate ϕ_j. Suppose that a new connection is accepted if and only if the resulting vector $n = (n_r, r \in \mathcal{R})$ satisfies

$$\sum_{r:j\in r} n_r\lambda_r < \phi_j - \epsilon$$

for each $j \in \mathcal{J}$. Show that, under the $M/M/1$ assumption, the mean packet delay at each queue is held below $1/\epsilon$.

6.1 Chernoff bound and Cramér's theorem

Our intuition for not wanting to use the peak rate for every user is based on the fact that, if the bandwidth requirements of the different users are independent, the total bandwidth requirement when we have many users should scale approximately as the mean. We will now attempt to understand the deviations: that is, with n independent users, how much larger can their total bandwidth requirement get than the sum of the mean rates?

Let $X_1, \ldots, X_n$ be independent identically distributed copies of a random variable X. Define $M(s) = \log \mathbb{E}e^{sX}$, the *log-moment generating function*. When we need to indicate dependence on the variable X, we write $M_X(s)$.

We shall make the assumption that $M(s)$ is finite for real s in an open neighbourhood of the origin, and thus that the moments of X are finite.

Now, for any random variable Y and any $s \geq 0$ we have the bound

$$\mathbb{P}(Y > 0) = \mathbb{E}[I[Y > 0]] \leq \mathbb{E}[e^{sY}]. \tag{6.1}$$

This is illustrated in Figure 6.2.

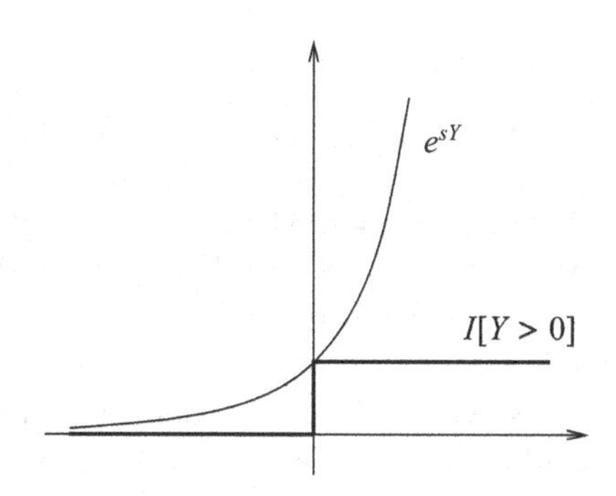

Figure 6.2 $I[Y > 0] \leq e^{sY}$.

Optimizing over s gives us the *Chernoff bound*:

$$\log \mathbb{P}(Y > 0) \leq \inf_{s \geq 0} \log \mathbb{E}[e^{sY}].$$

Now $M_{X_1 + \ldots + X_n}(s) = nM_X(s)$ and so

$$\log \mathbb{P}(X_1 + \ldots + X_n > 0) \leq n \inf_{s \geq 0} M(s).$$

We can easily put a number other than 0 into the right-hand side:

$$\begin{aligned} \log \mathbb{P}(X_1 + \ldots + X_n > nc) &= \log \mathbb{P}((X_1 - c) + \ldots + (X_n - c) > 0) \\ &\leq n \inf_{s \geq 0} [M_X(s) - cs], \end{aligned}$$

since $M_{X-c}(s) = M_X(s) - cs$. This gives us an upper bound on the probability that the sum $X_1 + \ldots + X_n$ is large; we show next that this upper bound is asymptotically exact as $n \to \infty$.

Remark 6.1 We have already assumed that $M(s)$ is finite for real s in an open neighbourhood of the origin. A consequence is that $M(s)$ is differentiable in the interior of the interval for which it is finite, with

$$M'(s) = \mathbb{E}(Xe^{sX})/\mathbb{E}(e^{sX}),$$

as we would expect, but we shall not prove this. (For a careful treatment, see Ganesh *et al.* (2004), Chapter 2.) We'll assume that the infimum in the following statement of Cramér's theorem is attained at a point in the interior of the interval for which $M(s)$ is finite.

Theorem 6.2 (Cramér's theorem) *Let $X_1, \ldots, X_n$ be independent identically distributed random variables with $\mathbb{E}[X_1] < c$, $\mathbb{P}(X_1 > c) > 0$, and suppose the log-moment generating function $M(s)$ satisfies the above assumptions. Then*

$$\lim_{n\to\infty} \frac{1}{n} \log \mathbb{P}(X_1 + \ldots + X_n > nc) = \inf_{s\geq 0} [M(s) - cs].$$

Proof We have shown that the right-hand side is an upper bound; let us now establish the lower bound. Let $s^* > 0$ be the value that achieves the infimum (it's greater than 0, since $M'(0) = \mathbb{E}[X_1] < c$).

We now use *tilting*. We define a new set of independent identically distributed random variables, $\tilde{X}_1, \ldots, \tilde{X}_n$, whose density is written in terms of the density of X_1. The tilted random variable $\tilde{X}_1$ will have mean $\mathbb{E}[\tilde{X}_1] = c$ and finite non-zero variance (Exercise 6.3), and so by the central limit theorem

$$\mathbb{P}\left(c < \frac{1}{n}(\tilde{X}_1 + \ldots + \tilde{X}_n) < c + \epsilon\right) \to \frac{1}{2} \quad \text{as } n \to \infty.$$

Since the densities of $\tilde{X}_i$ and X_i are related, this will allow us to bound from below the corresponding probability for X_i.

Let $f_X(\cdot)$ be the density of X_1. Define the density of the exponentially tilted random variable $\tilde{X}_1$ to be

$$f_{\tilde{X}}(x) = e^{-M(s^*)} e^{s^* x} f_{X_1}(x).$$

We need to check two things: first, that this is a density function, i.e. that it integrates to 1; and, second, that $\tilde{X}_1$ has mean c, as advertised. We first check that $f_{\tilde{X}}(\cdot)$ is a density function:

$$\int f_{\tilde{X}_1}(x)dx = e^{-M(s^*)} \int e^{s^* x} f_{X_1}(x)dx = e^{-M(s^*)} \mathbb{E}[e^{s^* X_1}] = 1$$

by the definition of M. We next check that $\tilde{X}_1$ has mean c. To do this, we recall that s^* was chosen as the point that minimized $M(s) - cs$; therefore, at s^*, the derivative of this function is 0, i.e. $c = M'(s^*)$. We now consider

$$\begin{aligned} \mathbb{E}[\tilde{X}_1] = \int x f_{\tilde{X}_1}(x)dx &= \int x e^{-M(s^*)} e^{s^* x} f_{X_1}(x)dx \\ &= \mathbb{E}[X_1 e^{s^* X_1 - M(s^*)}] = M'(s^*) = c \end{aligned}$$

as required.

We now bound the probability that $X_1 + \ldots + X_n > nc$ from below as follows:

$$\begin{aligned}
\mathbb{P}(X_1 + \ldots + X_n > nc) &> \mathbb{P}(nc < X_1 + \ldots + X_n < n(c+\epsilon)) \\
&= \int \cdots \int_{nc < x_1 + \ldots + x_n < n(c+\epsilon)} f_{X_1}(x_1)dx_1 \ldots f_{X_n}(x_n)dx_n \\
&= \int \cdots \int_{nc < x_1 + \ldots + x_n < n(c+\epsilon)} e^{M(s^*)} e^{-s^* x_1} f_{\tilde{X}_1}(x_1)dx_1 \ldots e^{M(s^*)} e^{-s^* x_n} f_{\tilde{X}_n}(x_n)dx_n \\
&= e^{nM(s^*)} \int \cdots \int_{nc < x_1 + \ldots + x_n < n(c+\epsilon)} e^{-s^*(x_1 + \ldots + x_n)} f_{\tilde{X}_1}(x_1)dx_1 \ldots f_{\tilde{X}_n}(x_n)dx_n \\
&\geq e^{nM(s^*)} e^{-s^* n(c+\epsilon)} \mathbb{P}(nc < \tilde{X}_1 + \ldots + \tilde{X}_n < n(c+\epsilon)).
\end{aligned}$$

The latter probability, as we discussed earlier, converges to $1/2$; so we conclude that

$$\frac{1}{n} \log \mathbb{P}(X_1 + \ldots + X_n > nc) \geq (M(s^*) - s^* c) - s^* \epsilon - \frac{1}{n} \log 2.$$

Letting $n \to \infty$ and $\epsilon \to 0$ gives us the result we wanted. □

Remark 6.3 An equivalent way of thinking about tilting is that we are changing the underlying probability measure from $\mathbb{P}$ to $\mathbb{Q}$, with $d\mathbb{Q}/d\mathbb{P} = e^{-M(s^*)} e^{s^* x}$. (This means that, for any measurable set S, $\mathbb{Q}(S) = \int_S e^{-M(s^*)} e^{s^* x} \, d\mathbb{P}(x)$.) Under $\mathbb{Q}$, we have $\mathbb{E}_{\mathbb{Q}} X_1 = c$. The random variable $\tilde{X}_1$ is simply X_1, and we compare $\mathbb{P}$- and $\mathbb{Q}$-probabilities of the event $\frac{1}{n}(X_1 + \ldots + X_n) \geq c$. This way of thinking about it doesn't require X_1 to have a density, which can be useful.

We know more from the proof than we have stated. Suppose the event $X_1 + \ldots + X_n > nc$ occurs. If n is large, then conditional on that event we can be reasonably certain of two things. First, the inequality is very close to equality. Second, the conditional distribution of $X_1, \ldots, X_n$ is very close to the distribution of n independent identically distributed random variables *sampled from the distribution of* $\tilde{X}_1$. That is, when the rare event occurs, we know rather precisely how it occurs; results from the theory of large deviations often have this flavour.

Exercises

Exercise 6.2 In this exercise, we establish some properties of the log-moment generating function, $M_X(s) = \log \mathbb{E}[e^{sX}]$. Check the following.

(1) If X and Y are independent random variables, $M_{X+Y}(s) = M_X(s) + M_Y(s)$.
(2) $M_X(s)$ is convex. [*Hint:* Check its derivative is increasing.]
(3) If X is a normal variable with mean λ and variance σ^2, then $M_X(s) = \lambda s + (\sigma s)^2/2$.

Exercise 6.3 We assumed that the infimum in the statement of Theorem 6.2 is attained in the interior of the interval for which $M(s)$ is finite. Check that this implies the tilted random variable $\tilde{X}_1$ has finite variance.

6.2 Effective bandwidth

Consider a single resource of capacity C, which is being shared by J types of connection. (For example, connections might be real-time audio or real-time video conference calls). Let n_j be the number of connections of type $j = 1, \ldots, J$. Let

$$S = \sum_{j=1}^{J} \sum_{i=1}^{n_j} X_{ji},$$

where X_{ji} are independent random variables, and where the distribution of X_{ji} can depend on j but not on i. Let

$$M_j(s) = \log \mathbb{E}[e^{sX_{ji}}].$$

We view X_{ji} as the bandwidth requirement of the ith connection of type j.

We should clearly expect to have $\mathbb{E}S < C$, and we will investigate what it takes to assure a small value for $\mathbb{P}(S > C)$.

Given C and information about the number and type of connections, the bound (6.1) implies that, for any $s \geq 0$,

$$\log \mathbb{P}(S > C) \leq \log \mathbb{E}[e^{s(S-C)}] = \sum_{j=1}^{J} n_j M_j(s) - sC.$$

In particular,

$$\inf_{s\geq 0} \left(\sum_{j=1}^{J} n_j M_j(s) - sC \right) \leq -\gamma \implies \mathbb{P}(S > C) \leq e^{-\gamma}. \qquad (6.2)$$

This is useful for deciding whether we can add another call of class k and still retain a service quality guarantee. Let

$$A = \left\{ n \in \mathbb{R}_+^J : \sum_{j=1}^{J} n_j M_j(s) - sC \leq -\gamma \text{ for some } s \geq 0 \right\}.$$

From the implication (6.2)

$$(n_1, \cdots, n_J) \in A \implies \mathbb{P}(S > C) \leq e^{-\gamma}.$$

Call A the *acceptance region*: a new connection can be accepted, without violating the service quality guarantee that $\mathbb{P}(S > C) \leq e^{-\gamma}$, if it leaves the vector n inside A.

Remark 6.4 The acceptance region A is conservative, since the implications are in one direction. Cramér's theorem shows that, for large values of C, n and γ, the converse holds, i.e. $\mathbb{P}(S > C) \leq e^{-\gamma}$ is approximately equivalent to $n \in A$ (Exercise 6.4).

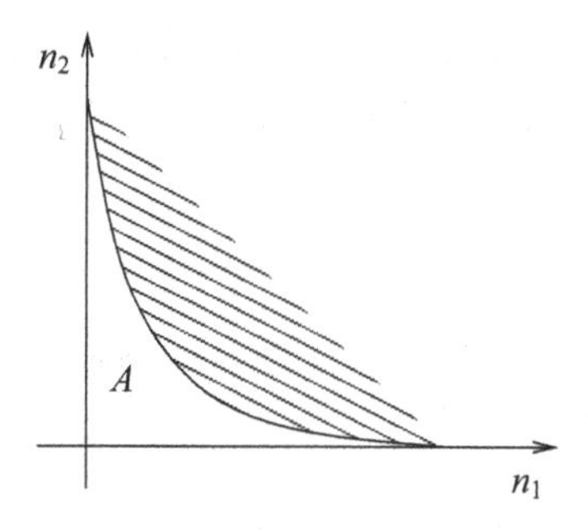

Figure 6.3 The convex complement of the acceptance region.

What does the acceptance region A look like? For each s, the inequality

$$\sum_{j=1}^{J} n_j M_j(s) - sC \leq -\gamma$$

defines a half-space of $\mathbb{R}^J$, and as s varies it indexes a family of half-spaces. The region A has a convex complement in $\mathbb{R}_+^J$ since this complement is just the intersection $\mathbb{R}_+^J$ with this family of half-spaces. Figure 6.3 illustrates the case $J = 2$. (As s varies, the boundary of the half-space rolls over the boundary of A, typically forming a tangent plane.)

Define

$$\alpha_j(s) = \frac{M_j(s)}{s} = \frac{1}{s}\log \mathbb{E}[e^{sX_{ji}}].$$

Rewriting the inequality, we see that if

$$\sum_{j=1}^{J} n_j\alpha_j(s) \le C - \frac{\gamma}{s} \tag{6.3}$$

we are guaranteed to have $\mathbb{P}(S > C) \le e^{-\gamma}$. But how should we choose s? If we know the typical operating region, the natural choice is to look for a point n^* on the boundary of A corresponding to the typical traffic mix, and then choose s to attain the infimum in (6.2) for $n = n^*$, as illustrated in Figure 6.4.

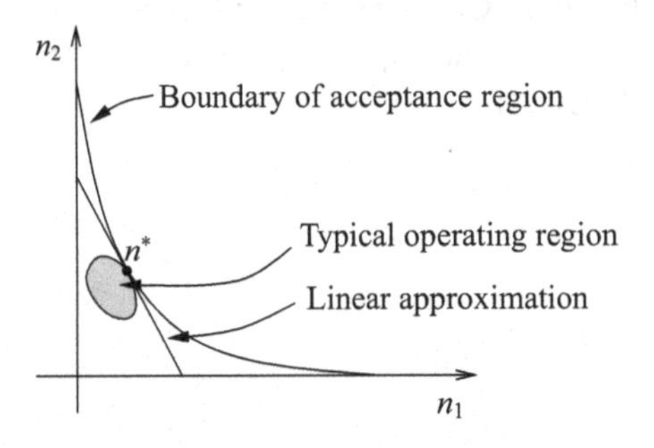

Figure 6.4 Approximating the acceptance region by a half-space bounded by the tangent at n^*.

We thus have a simple admission control: accept a new connection if it leaves the inequality (6.3) satisfied. We call $\alpha_j(s)$ the *effective bandwidth* of a source of class j. The admission control simply adds the effective bandwidth of a new request to the effective bandwidths of connections already in progress and accepts the new request if the sum satisfies a bound.

If a similar admission control is enforced at each resource of a network, the network will behave as a loss network. Note that the effective bandwidth of a connection may vary over the resources of the network (since different resources may choose different values of s, depending on their mix of traffic), just as in our earlier model of a loss network the requirements A_{jr} may vary over resources j for a given route r. Techniques familiar from loss networks can be used, for example trunk reservation to control relative priorities for different types of connection (Exercise 3.12).

Remark 6.5 The admission criterion (6.3) is conservative for two reasons. First, the acceptance region A is conservative, as noted in Remark 6.4. Second, the boundary of A may not be well approximated by a tangent plane; we explore this next.

Example 6.6 (Gaussian case) How well approximated is the acceptance region A by the linear constraint (6.3)? Some insight can be obtained from the Gaussian case, where explicit calculations are easy. Suppose that

$$\alpha_j(s) = \lambda_j + \frac{s\sigma_j^2}{2},$$

corresponding to a normally distributed load with mean λ_j and variance σ_j^2. Then (Exercise 6.7) the acceptance region is given by

$$A = \left\{ n : \sum_j n_j \lambda_j + \left(2\gamma \sum_j n_j \sigma_j^2 \right)^{1/2} \leq C \right\},$$

and the tangent plane at a point n^* on the boundary of A is of the form (6.3), with

$$\alpha_j(s) = \lambda_j + \gamma \frac{\sigma_j^2}{C(1-\rho^*)},$$

where $\rho^* = \sum_j n_j^* \lambda_j / C$, the traffic intensity associated with the point n^*.

Thus the effective bandwidths $\alpha_j(s)$ will be relatively insensitive to the traffic mix n^* provided $(1-\rho^*)^{-1}$ does not vary too much with n^*, or, equivalently, provided the traffic intensity is not too close to 1 on the boundary of A. Now, traffic intensity very close to 1 is only sustainable if the variance of the load is very small: this might arise near the n_k axis if traffic type k is nearly deterministic.

Exercises

Exercise 6.4 Use Cramér's theorem to show that

$$\lim_{N\to\infty} \frac{1}{N} \log \mathbb{P}\left(\sum_{j=1}^{J} \sum_{i=1}^{n_j N} X_{ji} > CN \right) = \inf_{s\geq 0} \left[\sum_{j=1}^{J} n_j M_j(s) - sC \right].$$

In this sense, the converse to the implication (6.2) holds as the number of sources increases and the tail probability decreases.

Exercise 6.5 In this exercise, we explore properties of the effective bandwidth formula

$$\alpha(s) = \frac{1}{s} \log \mathbb{E}[e^{sX}].$$

Show that $\alpha(s)$ is increasing in s over the range $s \in (0, \infty)$ and that it lies between the mean and the peak:

$$\mathbb{E}X \leq \alpha(s) \leq \sup\{x : \mathbb{P}(X > x) > 0\}.$$

[*Hint:* Jensen's inequality.]

Further, show that

$$\lim_{s\to 0} \alpha(s) = \mathbb{E}X \quad \text{and} \quad \lim_{s\to\infty} \alpha(s) = \text{ess sup}\, X.$$

Exercise 6.6 If X is a random variable with $\mathbb{P}(X = h) = p = 1-\mathbb{P}(X = 0)$, show that its effective bandwidth is given by

$$\frac{1}{s} \log(pe^{sh} + 1 - p).$$

Exercise 6.7 Consider the Gaussian case, Example 6.6. Show that the infimum in relation (6.2) with $n = n^*$ is attained at

$$s = \frac{C - \sum_j n_j^* \lambda_j}{\sum_j n_j^* \sigma_j^2},$$

and hence deduce the expressions for A and for a tangent plane given in Example 6.6.

Check that, with this value of s, the right-hand side of inequality (6.3) can be written as

$$C - \gamma \frac{\sum_j n_j^* \sigma_j^2}{C(1-\rho^*)} = C - \gamma \frac{\text{variance of load}}{\text{mean free capacity}}.$$

For the Gaussian case, we can compute a necessary and sufficient condition on n such that $\mathbb{P}(S > C) \leq e^{-\gamma}$; show that it takes the form

$$\sum_j n_j \lambda_j + \phi \left(\sum_j n_j \sigma_j^2 \right)^{1/2} \leq C,$$

where $\phi = \Phi^{-1}(1 - e^{-\gamma})$ and Φ is the standard normal distribution function.

6.3 Large deviations for a queue with many sources

Next we consider a model that includes a buffer: this allows the total bandwidth requirements of all the connections to exceed the capacity for a while. The challenge is to work out for how long!

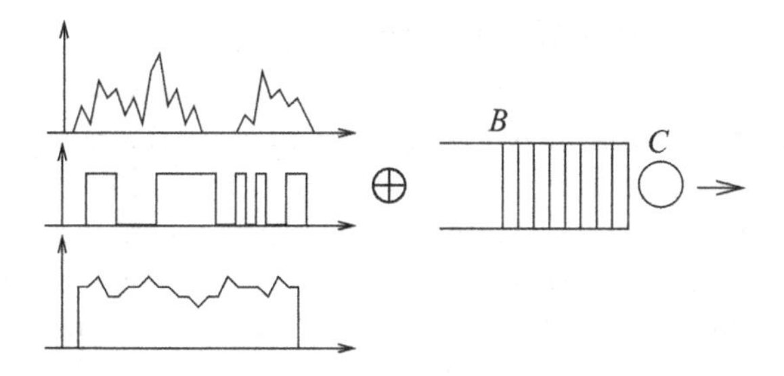

Figure 6.5 Three users (of three different types) feeding through a buffer.

We consider the following model. There is a potentially infinite queue in which unprocessed work resides. As long as the queue is positive, the work is removed from it at rate C; when the queue is empty, work is processed at rate C or as quickly as it arrives. The system experiences "pain" (packets dropped, or held up at earlier queues, etc.) when the queue level exceeds B; this event is called buffer overflow. We would like to give a description of the set of vectors n (giving the number of connections of each type) for which the probability of buffer overflow is small.

Because the queue introduces a time component into the picture, we need slightly more notation. Let $X_{ji}[t_1, t_2]$ be the workload arriving from the ith source of type j during the interval $[t_1, t_2]$. We assume that the random processes (X_{ji}) are independent, that the distributions may depend upon the type j but not upon i, and that each process has *stationary increments*, i.e. the distribution of $X_{ji}[t_1, t_2]$ depends on t_1, t_2 only through $t_2 - t_1$.

Next, we give some simple examples of processes with stationary increments.

Example 6.7 (Constant-rate sources) Suppose $X_{ji}[t_1, t_2] = (t_2 - t_1)X_{ji}$, where X_{ji} is a non-negative random variable. That is, each connection has a constant (but random) bandwidth requirement, whose distribution depends on the user type j. This essentially reproduces the bufferless model of Section 6.2: if $\sum X_{ji} < C$, buffer overflow never occurs; if $\sum X_{ji} > C$, buffer overflow will definitely occur from a finite time onwards, no matter how large the buffer.

Example 6.8 ($M/G/1$ queue) Suppose $X_{ji}[0,t] = \sum_{k=1}^{N(t)} Z_k$, where $N(t)$ is a Poisson process and Z_k are independent identically distributed random variables. This models the arrival of work at an $M/G/1$ queue: the interarrival times are exponential (hence the first M), and the service requirements of different "customers" are independent but have a general, not necessarily exponential, distribution (hence the G).

Remark 6.9 We might be more interested in the case of a truly finite buffer, in which work that arrives when the buffer is full is never processed. We might expect that buffer overflow would occur more rarely in this system, as its queue should be shorter. In the regime where the buffer rarely overflows (this is the regime we would like to be in!), the overflows that do occur will be small; so there is not much difference between dropping this work and keeping it. This can be made precise; see Ganesh *et al.* (2004), Chapter 7.

Suppose the number of connections of type j is n_j, let $n = (n_1, \cdots, n_J)$, and let the workload arriving at the queue over the interval $[t_1, t_2]$ be

$$S[t_1,t_2] = \sum_{j=1}^{J}\sum_{i=1}^{n_j} X_{ji}[t_1,t_2].$$

We suppose the queue has been evolving from time $-\infty$ and look at the queue size at time 0; because the arrival processes X_{ji} are time homogenous, this will have the same distribution as the queue size at any other time t. The queue size at time 0 is the random variable

$$Q(0) = \sup_{0\le\tau<\infty} \left(S[-\tau,0] - C\tau\right). \tag{6.4}$$

To see this, note that the queue at time 0 must be at least $S[-\tau,0] - C\tau$ for any τ, because it cannot clear work any faster than that; that the queue size is indeed given by (6.4) is shown in Exercise 6.8. There we shall also see that the random variable $-\tau$ is the time the queue was last empty before time 0 (more exactly, in the case of non-uniqueness, the random variable $-\tau$ is such that $Q(-\tau) = 0$ and the queue drains at the full rate C over $(-\tau,0)$).

The random variable $Q(0)$ may be infinite. For example, with constant-rate sources of total rate S the queue size is infinite with probability $\mathbb{P}(S > C)$, precisely the object of study in Section 6.2, as we noted in Example 6.7.

Let $L(C,B,n) = \mathbb{P}(Q(0) > B)$; this is the probability we would like to control. We are going to look at this in the regime where capacity, number of connections, and buffer size are all large, while the number of connection types is fixed.

Theorem 6.10

$$\lim_{N\to\infty}\frac{1}{N}\log L(CN, BN, nN) = \sup_{t\geq 0}\inf_{s\geq 0}\left[st\sum_{j\in J} n_j\alpha_j(s,t) - s(B+Ct)\right],$$

where

$$\alpha_j(s,t) = \frac{1}{st}\log \mathbb{E}e^{sX_{ji}[0,t]}. \tag{6.5}$$

Sketch of proof The probability

$$\mathbb{P}(Q^N(0) > BN) = \mathbb{P}\left(S^N[-t,0] > (B+Ct)N \text{ for some } t\right)$$

is bounded from below by $\mathbb{P}(S^N[-t,0] > (B+Ct)N)$, for any fixed t, and from above by $\sum_{t\geq 0}\mathbb{P}(S^N[-t,0] > (B+Ct)N)$ (at least in discrete time).

For any fixed t, Cramér's theorem gives

$$\frac{1}{N}\log\mathbb{P}\left(\sum_{j=1}^{J}\sum_{i=1}^{n_jN} X_{ji}[-t,0] > (B+Ct)N\right)$$
$$\to \inf_s\left[st\sum_{j=1}^{J} n_j\alpha_j(s,t) - s(B+Ct)\right].$$

This gives us the lower bound directly.

The upper bound can be derived by noting that the terms decay exponentially in N, and therefore we can ignore all but the largest term. Making this precise takes some work, but see Exercise 6.9 for a start. □

Suppose that s^* and t^* are the extremizing parameter values in Theorem 6.10. The sketched proof allows an interpretation of these values. We see that if buffer overflow occurs, then with high probability this happens because, over a time period t^*, the total work that arrived into the system was greater than $B + Ct^*$. Thus, t^* is the typical time scale for buffer overflow. The most likely trajectory for the amount of work in the buffer over the time scale $[-t^*, 0]$ (where $t = 0$ is the time of the buffer overflow event) depends on the temporal correlation structure of the processes X_{ji}. Figure 6.6 sketches some possibilities.

The parameter s^* is the exponential tilt for the distribution of S, and for each of the X_{ji}, that makes the sum S over the time period t^* likely to be $B + Ct^*$.

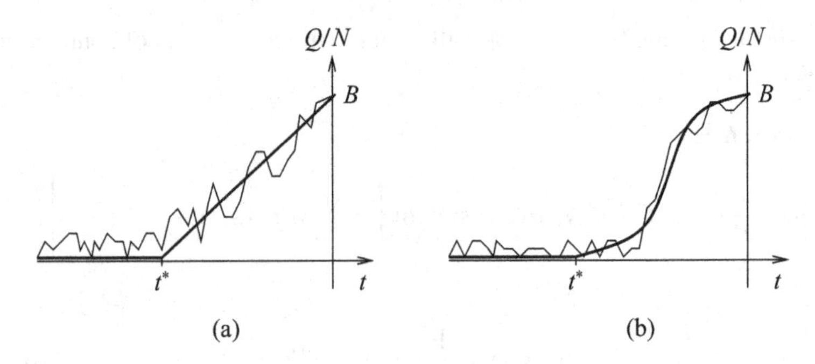

Figure 6.6 Most likely trajectories leading to buffer overflow. The thick lines show the limit as $N \to \infty$; the thin lines depict possible instances for finite N. (a) If S has independent increments, the most likely trajectory is linear. (b) Shown is the most likely trajectory if traffic is described by a fractional Brownian motion with Hurst parameter $> 1/2$; this is a process exhibiting positive correlations over time and long-range dependence.

An acceptance region can be defined as in Section 6.2. Although its complement may no longer be convex, the expressions (6.5) do define the tangent plane at a typical boundary point, and hence have interpretations as effective bandwidths.

Remark 6.11 The notion of an effective bandwidth can be developed in various other frameworks. For example, Exercise 6.10 develops a linearly bounded acceptance region for a model where the log-moment generating function may not be defined (other than at 0). But the large deviations framework sketched in this section does give a broad concept of effective bandwidth that takes into account the statistical characteristics of a connection over different time and space scales.

Exercises

Exercise 6.8 In this exercise we derive equation (6.4) for the queue size. We suppose the queue has been evolving from time $-\infty$. If the queue is draining at the full rate C for the entire period $(-\infty, 0)$, there may be an ambiguity in the definition of the queue size. This might occur, for example, with constant-rate sources such that $S = C$, since then the queue size is a fixed constant for all t, which we can interpret as the initial queue size at time $-\infty$. We assume the queue size is 0 at least once in the past, possibly in the limit as $t \to -\infty$.

Suppose that $\tau < \infty$ achieves the supremum in equation (6.4).

- Show that $S[-\tau, -t] \geq C(\tau - t)$ for all $0 \leq t \leq \tau$, and deduce that the queue must be draining at the full rate C for the entire period $[-\tau, 0]$.
- Show that $S[-t, -\tau] \leq C(t - \tau)$ for all $t \geq \tau$, and deduce that $Q(-t) \geq Q(-\tau)$ for all $t \geq \tau$.
- Deduce that $Q(-\tau) = 0$, and hence equation (6.4).

Finally obtain equation (6.4) in the case where the supremum attained is finite, but is only approached as $\tau \to \infty$.

Exercise 6.9 (Principle of the largest term) Consider two sequences a_N, b_N, with

$$\frac{1}{N} \log a_N \to a, \qquad \frac{1}{N} \log b_N \to b.$$

If $a \geq b$, show that

$$\frac{1}{N} \log(a_N + b_N) \to a.$$

Conclude that, for finitely many sequences $p^N(0), \ldots, p^N(T)$ with limits $p(0), \ldots, p(T)$,

$$\frac{1}{N} \log\left(\sum_{t=0}^{T} p^N(t)\right) \to \max_{0 \leq t \leq T} p(t) \qquad \text{as } N \to \infty.$$

That is, the growth of a finite sum is governed by the largest growth rate of any term in it; this is known as the principle of the largest term.

In the proof of Theorem 6.10, we need to bound $\sum_{t \geq 0} \mathbb{P}(X^N[-t, 0] > (B + Ct)N)$, which involves extending this argument to a countable sum.

Exercise 6.10 The expected amount of work in an $M/G/1$ queue, Example 6.8, is

$$\mathbb{E}Q = \frac{\nu(\mu^2 + \sigma^2)}{2(C - \nu\mu)},$$

where ν is the rate of the Poisson arrival process and μ and σ^2 are the mean and variance, respectively, of the distribution G (from the Pollaczek–Khinchine formula). Now suppose that

$$G(x) = \sum_{j=1}^{J} p_j G_j(x), \quad \nu = \sum_{j=1}^{J} \nu_j n_j, \quad p_j = \frac{\nu_j n_j}{\nu},$$

corresponding to n_j sources of type j, with ν_j the arrival rate and G_j the service requirement distribution of type j. Let μ_j and σ_j^2 be the mean and variance, respectively, of the distribution G_j. Show that the service quality guarantee $\mathbb{E}Q \leq B$ is satisfied if and only if

$$\sum_{j=1}^{J} \alpha_j n_j \leq C,$$

where

$$\alpha_j = \nu_j \left[\mu_j + \frac{1}{2B}(\mu_j^2 + \sigma_j^2) \right].$$

6.4 Further reading

Ganesh *et al.* (2004) give a systematic account of how large deviations theory can be applied to queueing problems, and give extensive references to the earlier papers. For a more detailed development of the examples given in this chapter, the reader is referred to Kelly (1996) and to Courcoubetis and Weber (2003). For a treatment of effective bandwidths within a more general setting of performance guarantees, see Chang (2000). Mazumdar (2010) emphasizes the role of effective bandwidths as mappings from queueing level phenomena to loss network models, and provides a concise mathematical treatment of a number of topics from this book.

The effective bandwidth of a connection is a measure of the connection's consumption of network resources, and we might expect it to be relevant in the calculation of a charge: see Songhurst (1999) for a detailed discussion of usage-based charging schemes, and Courcoubetis and Weber (2003) for a broad overview of the complex subject of pricing communication networks.

After a flow has passed through a resource, its characteristics may change: Wischik (1999) considers a model where the relevant statistical characteristics of a flow of traffic are preserved by passage through a resource, in the limit where the number of inputs to that resource increases.

Part III

7

Internet congestion control

In Chapter 6 we looked at organizing a network of users with time-varying bandwidth requirements into something that resembles a loss network, by finding a notion of effective bandwidth. In this chapter we consider an alternative approach, where the users actively participate in the sharing of the network resources.

7.1 Control of elastic network flows

How should available resources be shared between competing streams of elastic traffic? There are at least two aspects to this question. First, why would one allocation of resources be preferred to another? Second, what control mechanisms could be implemented in a network to achieve any preferred allocation? To make progress with these questions, we need to fix some notation and define some terms.

Consider a network with a set $\mathcal{J}$ of *resources*, and let C_j be the finite capacity of resource j. (We occasionally refer to the resources as *links*.) Let a *route* r be a non-empty subset of $\mathcal{J}$, and write $\mathcal{R}$ for the set of possible routes. As before, we use the *link-route incidence matrix* A to indicate which resources belong to which routes. A is a $|\mathcal{J}| \times |\mathcal{R}|$ matrix with $A_{jr} = 1$ if $j \in r$, and $A_{jr} = 0$ otherwise.

We will identify a user with a route. For example, when Elena at the University of Michigan downloads the web page at `www.google.com`, the corresponding "user" (route) is sending data along 16 links, including the link from Elena's laptop to the nearest wireless access point. If Elena also downloads the web page `www.cam.ac.uk`, this (different!) "user" is using 18 links; five of the links are common with the route to Google.

Suppose that, if a rate x_r is allocated to user r, this has a utility $U_r(x_r)$. We formalize the notion of elastic traffic by assuming that the *utility function* $U_r(\cdot)$ is increasing and concave, and that our objective is to maximize the sum of user utilities. For convenience, we also assume that $U_r(\cdot)$ is strictly concave, continuously differentiable on $(0, \infty)$, and satisfies $U_r'(0) = \infty$ and $U_r'(\infty) = 0$.

Remark 7.1 Elastic traffic, i.e. concave utility functions, captures the notion of users preferring to share. The aggregate utility of two users both experiencing a medium download rate is higher than if one of them got a very high rate and the other a very low one. If more users come into the system, it is better for existing users to decrease their rates, so that everyone has a chance to use the network. This is not true of all types of traffic. For real-time voice communication, the utility function might have the form shown on the right in Figure 7.1 if speech becomes unintelligible at low rates. In this case, users may require a certain set of resources in order to proceed, and attempting to share would decrease aggregate utility. If there are too many users then some should be served and some not; in a loss network the implicit randomization is effected by the admission control mechanism.

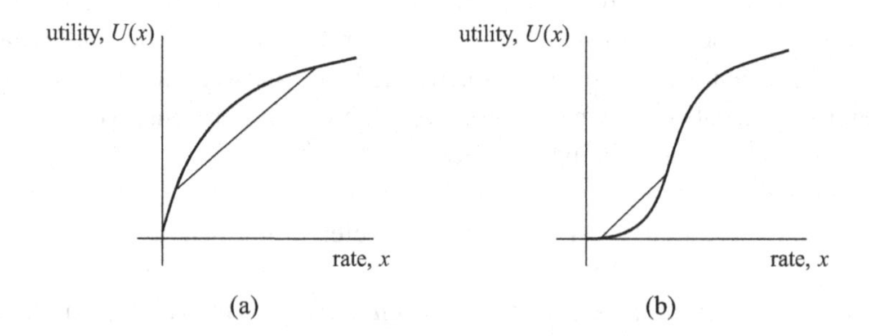

Figure 7.1 Elastic and inelastic demand. (a) Elastic traffic, prefers to share; (b) inelastic traffic, prefers to randomize.

Let $C = (C_j, j \in \mathcal{J})$ be the vector of capacities, let $U = (U_r(\cdot), r \in \mathcal{R})$ be the collection of user utilities, and let $x = (x_r, r \in \mathcal{R})$ be an allocation of flow rates. The feasible region is then given by $x \geq 0, Ax \leq C$, and under the model we have described the system optimal rates solve the following problem.

SYSTEM(U, A, C):

$$\begin{aligned} &\text{maximize} && \sum_{r \in \mathcal{R}} U_r(x_r) \\ &\text{subject to} && Ax \leq C \\ &\text{over} && x \geq 0. \end{aligned} \tag{7.1}$$

While this optimization problem is mathematically tractable (with an objective function that is strictly concave and a convex feasible region), it involves utilities U that are unlikely to be known by the network; further, even the matrix A describing the network topology is not likely to be known in its entirety at any single point in the network. We address the issue of unknown utilities first, by decomposing the optimization problem into simpler problems. We leave the second issue for later sections.

Suppose that user r may choose an amount to pay per unit time, w_r, and will then receive in return a flow rate x_r proportional to w_r, where $1/\lambda_r$ is the known constant of proportionality. Then the utility maximization problem for user r is as follows.

USER$_r(U_r; \lambda_r)$:

$$\begin{aligned} &\text{maximize} && U_r\left(\frac{w_r}{\lambda_r}\right) - w_r \\ &\text{over} && w_r \geq 0. \end{aligned} \tag{7.2}$$

Here we could interpret w_r as user r's budget and λ_r as the price per unit of flow; note that no capacity constraints are involved.

Suppose next that the network knows the vector $w = (w_r, r \in \mathcal{R})$ and solves a variant of the SYSTEM problem using a particular set of utility functions, $U_r(x_r) = w_r \log x_r, r \in \mathcal{R}$.

NETWORK$(A, C; w)$:

$$\begin{aligned} &\text{maximize} && \sum_{r \in \mathcal{R}} w_r \log x_r \\ &\text{subject to} && Ax \leq C \\ &\text{over} && x \geq 0. \end{aligned} \tag{7.3}$$

We can find a pleasing relationship between these various problems, summarized in the following theorem.

Theorem 7.2 (Problem decomposition) *There exist vectors* $\lambda = (\lambda_r, r \in \mathcal{R})$, $w = (w_r, r \in \mathcal{R})$ *and* $x = (x_r, r \in \mathcal{R})$ *such that*

- $w_r = \lambda_r x_r$ *for all* $r \in \mathcal{R}$,
- w_r *solves* USER$_r(U_r; \lambda_r)$ *for all* $r \in \mathcal{R}$,
- x *solves* NETWORK$(A, C; w)$.

Moreover, the vector x *then also solves* SYSTEM(U, A, C).

Remark 7.3 Thus the system problem can be solved by simultaneously solving the network and user problems. (In later sections, we shall address the issue of how to solve the problems simultaneously, but you might anticipate that some form of iterative procedure will be involved.)

There are many other ways to decompose the SYSTEM problem. For example, we might pick a version of the NETWORK problem that uses a different set of "placeholder" utilities, instead of the family $w_r \log(\cdot)$; we consider a more general class in Section 8.2. As we shall see, the logarithmic utility function occupies a privileged position within a natural family of scale-invariant utility functions, and it corresponds to a particular notion of fairness.

Proof Note first of all that there exists a unique optimum for the SYSTEM problem, because we are optimizing a strictly concave function over a closed convex set; and the optimum is interior to the positive orthant since $U_r'(0) = \infty$. Our goal is to show that this optimum is aligned with an optimum for each of the NETWORK and the USER problems.

The Lagrangian for SYSTEM is

$$L_{\text{SYSTEM}}(x, z; \mu) = \sum_{r \in \mathcal{R}} U_r(x_r) + \mu^T (C - Ax - z),$$

where $z = (z_j, j \in \mathcal{J})$ is a vector of slack variables, and $\mu = (\mu_j, j \in \mathcal{J})$ is a vector of Lagrange multipliers, for the inequality constraints. By the strong Lagrangian principle (Appendix C), there exists a vector μ such that the Lagrangian is maximized at a pair x, z that solve the original problem, SYSTEM.

In order to maximize the Lagrangian over x_r, we differentiate:

$$\frac{\partial}{\partial x_r} L_{\text{SYSTEM}}(x, z; \mu) = U_r'(x_r) - \sum_{j \in r} \mu_j,$$

since differentiating the term $\mu^T Ax$ will leave only those μ_j with $j \in r$.

The Lagrangian for NETWORK is

$$L_{\text{NETWORK}}(x, z; \tilde{\mu}) = \sum_{r \in \mathcal{R}} w_r \log x_r + \tilde{\mu}^T (C - Ax - z),$$

and on differentiating with respect to x_r we obtain

$$\frac{\partial}{\partial x_r} L_{\text{NETWORK}}(x, z; \tilde{\mu}) = \frac{w_r}{x_r} - \sum_{j \in r} \tilde{\mu}_j.$$

(There's no reason at this point to believe that μ and $\tilde{\mu}$ have anything to do with each other; they are simply the Lagrange multipliers for two different optimization problems.)

Finally, for USER the objective function is $U_r(w_r/\lambda_r) - w_r$, and

$$\frac{\partial}{\partial w_r} \left[U_r \left(\frac{w_r}{\lambda_r} \right) - w_r \right] = \frac{1}{\lambda_r} U_r' \left(\frac{w_r}{\lambda_r} \right) - 1.$$

Suppose now that user r always picks her budget w_r to optimize the USER objective function. Then we must have

$$U_r' \left(\frac{w_r}{\lambda_r} \right) = \lambda_r.$$

Therefore, if we pick

$$\tilde{\mu}_j = \mu_j, \quad \lambda_r = \sum_{j \in r} \mu_j, \quad w_r = \lambda_r (U_r')^{-1}(\lambda_r), \quad x_r = \frac{w_r}{\lambda_r},$$

the NETWORK and SYSTEM Lagrangians will have the same derivatives. Thus, if μ_j are optimal dual variables for the SYSTEM problem, this choice of the remaining variables simultaneously solves all three problems and establishes the existence of the vectors λ, w and x satisfying conditions 1, 2 and 3 of the theorem. Conversely, suppose vectors λ, w and x satisfy conditions 1, 2 and 3: then x solves the NETWORK problem, and so can be written in terms of $\tilde{\mu}$, and setting $\mu = \tilde{\mu}$ shows that x also solves the SYSTEM problem. □

We can identify μ_j as the *shadow price* per unit flow through resource j. The price for a unit of flow along route r is λ_r, and it decomposes as a sum over the resources involved.

Next we determine the dual problem. The value of the NETWORK Lagrangian is

$$\max_{x,z \geq 0} L_{\text{NETWORK}}(x, z; \mu) = \sum_{r \in \mathcal{R}} w_r \log \frac{w_r}{\sum_{j \in r} \mu_j} + \mu^T C.$$

If we subtract the constant $\sum_r w_r \log w_r$, and change sign, we obtain the following optimization problem.

DUAL$(A, C; w)$:

$$\text{maximize} \quad \sum_{r \in \mathcal{R}} w_r \log\left(\sum_{j \in r} \mu_j\right) - \sum_{j \in J} \mu_j C_j$$
$$\text{over} \quad \mu \geq 0.$$

This rephrases the NETWORK problem as a problem of setting up prices μ_j per unit of flow through each resource so as to optimize a certain revenue function.

The treatment in this section suggests that one way to solve the resource allocation problem SYSTEM without knowing the individual user utilities would be to charge the users for access to the network, and allow users to choose the amounts they pay. But, at least in the early days of the Internet, charging was both impractical and likely to ossify a rapidly developing technology. In Section 7.2 we discuss how one might compare alternative resource allocations using the ideas in this section, but without resorting to charging.

Exercises

Exercise 7.1 In the USER problem, user r acts as a price-taker: she does not anticipate the consequences of her choice of w_r upon the price λ_r. In this exercise, we explore the effect of more strategic behaviour.

Consider a single link of capacity C, shared between users in $\mathcal{R}$. Suppose that user r chooses w_r: show that, under the solution to NETWORK, user r receives a rate

$$\frac{w_r}{\sum_{s \in \mathcal{R}} w_s} C.$$

Now suppose user r anticipates that this will be the outcome of the choices $w = (w_s, s \in \mathcal{R})$ and attempts to choose w_r to maximize

$$U_r\left(\frac{w_r}{\sum_{s \in \mathcal{R}} w_s} C\right) - w_r$$

over $w_r > 0$, for given values of w_s, $s \neq r$. Show that there exists a vector w that solves this problem simultaneously for each $r \in \mathcal{R}$, and that it is the unique solution to the problem

$$\begin{array}{ll} \text{maximize} & \sum_{r \in \mathcal{R}} \hat{U}_r(x_r) \\ \text{subject to} & \sum_{r \in \mathcal{R}} x_r \le C \\ \text{over} & x_r \ge 0, \end{array}$$

where

$$\hat{U}_r(x_r) = \left(1 - \frac{x_r}{C}\right) U_r(x_r) + \frac{x_r}{C}\left(\frac{1}{x_r}\int_0^{x_r} U_r(y)dy\right).$$

Thus the choices of price-anticipating users will not in general maximize the sum of user utilities, although, if many users share a link and $x_r \ll C$, the effect is likely to be small.

Exercise 7.2 In this exercise, we extend the model of this section to include routing.

Let $s \in \mathcal{S}$ now label a user (perhaps a source–destination pair), and suppose s is identified with a subset of $\mathcal{R}$, namely the routes available to serve the user s. Define an incidence matrix H as in Section 4.2.2, by setting $H_{sr} = 1$ if $r \in s$ and $H_{sr} = 0$ otherwise. Thus if y_r is the flow on route r and $y = (y_r, r \in \mathcal{R})$, then $x = Hy$ gives the aggregate flows achieved by each user $s \in S$. Let $\text{SYSTEM}(U, H, A, C)$ be the problem

$$\begin{array}{lll} \text{maximize} & \sum_{s \in \mathcal{S}} U_s(x_s) & \\ \text{subject to} & x = Hy, & Ay \le C, \\ \text{over} & x, y \ge 0, & \end{array}$$

and let $\text{NETWORK}(H, A, C; w)$ be the problem

$$\begin{array}{lll} \text{maximize} & \sum_{r \in \mathcal{R}} w_s \log x_s & \\ \text{subject to} & x = Hy, & Ay \le C, \\ \text{over} & x, y \ge 0. & \end{array}$$

Show that there exist vectors $\lambda = (\lambda_s, s \in \mathcal{S})$, $w = (w_s, s \in \mathcal{S})$ and $x = (x_s, s \in \mathcal{S})$ satisfying $w_s = \lambda_s x_s$ for $s \in \mathcal{S}$, such that w_s solves $\text{USER}_s(U_s; \lambda_s)$ for $s \in \mathcal{S}$ and x solves $\text{NETWORK}(H, A, C; w)$; show further that x is then the unique vector with the property that there exists a vector y such that (x, y) solves $\text{SYSTEM}(U, H, A, C)$.

Exercise 7.3 Requests to view a web page arrive as a Poisson process of rate C. Each time a page is viewed, an advert is displayed within the page. The advert is selected randomly from a pool of $\mathcal{R}$ adverts; advert $r \in \mathcal{R}$ is selected with probability $w_r / \sum_{s \in \mathcal{R}} w_s$, where w_r is the payment per unit time (e.g. per day) made by advertiser r. Advert r, when displayed, is clicked upon by a viewer of the page with probability $1/A_r$. Advertiser r does not observe C, A_s for $s \in \mathcal{R}$, or w_s for $s \neq r$. Assume that advertiser r *does* observe

$$x_r \equiv \frac{w_r}{\sum_{s \in \mathcal{R}} w_s} \cdot \frac{C}{A_r};$$

this is a reasonable assumption, since the number of clicks on advert r has mean x_r under the above model. Advertiser r can thus deduce

$$\lambda_r \equiv \frac{w_r}{x_r},$$

the amount she has paid per click.

Show that, under the usual assumptions on $U_r(\cdot), r \in \mathcal{R}$, there exist vectors $\lambda = (\lambda_r, r \in \mathcal{R})$, $w = (w_r, r \in \mathcal{R})$ and $x = (x_r, r \in \mathcal{R})$ such that the above two identities are satisfied and such that, for each $r \in \mathcal{R}$, w_r maximizes $U_r(w_r/\lambda_r) - w_r$. Show that the vector x then also maximizes $\sum_{r \in \mathcal{R}} U_r(x_r)$ over all x satisfying $\sum_{r \in \mathcal{R}} A_r x_r = C$.

7.2 Notions of fairness

We would like the allocation of flow rates to the users to be fair in some sense, but what do we mean by this? There are several widely used notions of fairness; we will formulate a few here. Later, in Section 8.2, we encounter a generalized notion of fairness that encompasses all of these as special cases.

Max-min fairness. We say that an allocation $x = (x_r, r \in \mathcal{R})$ is *max-min fair* if it is feasible (i.e. $x \geq 0$ and $Ax \leq C$) and, for any other feasible allocation y,

$$\exists r : y_r > x_r \implies \exists s : y_s < x_s \leq x_r.$$

That is, in order for user r to benefit, someone else (user s) who was worse off than r needs to get hurt. The compactness and convexity of the feasible region imply that such a vector x exists and is unique (check this!). The term "max-min" comes from the fact that we're maximizing the minimal amount that anyone is getting: i.e. the max-min fair allocation will give a solution to

$$\max_{x \text{ feasible}} (\min_r x_r).$$

Actually, the max-min fair allocation will give more: once we've fixed the smallest amount that anyone is getting, we then maximize the second-smallest, then the third-smallest, and so on.

The concept of max-min fairness has been much discussed by political philosophers. In *A Theory of Justice*, Rawls (1971) proposes that, if one were to design society from scratch, without knowing where in it one might end up, one would want it to be max-min fair. However, for our purposes, it seems a bit extreme, because it places such overwhelming emphasis on maximizing the lowest rate.

Proportional fairness. We say an allocation $x = (x_r, r \in \mathcal{R})$ is *proportionally fair* if it is feasible, and if for any other feasible allocation y the aggregate of proportional changes is non-positive:

$$\sum_r \frac{y_r - x_r}{x_r} \leq 0.$$

This still places importance on increasing small flows (for which the denominator is smaller), but it is not quite as overwhelming.

There may be a priori reasons why some users should be given more weight than others: perhaps a route is providing a flow of value to a number of individuals (these lecture notes are being downloaded to a class), or different users invested different amounts initially to construct the network. Let $w = (w_r, r \in \mathcal{R})$ be a vector of weights: then $x = (x_r, r \in \mathcal{R})$ is *weighted proportionally fair* if it is feasible, and if for any other feasible vector y

$$\sum_r w_r \frac{y_r - x_r}{x_r} \leq 0. \tag{7.4}$$

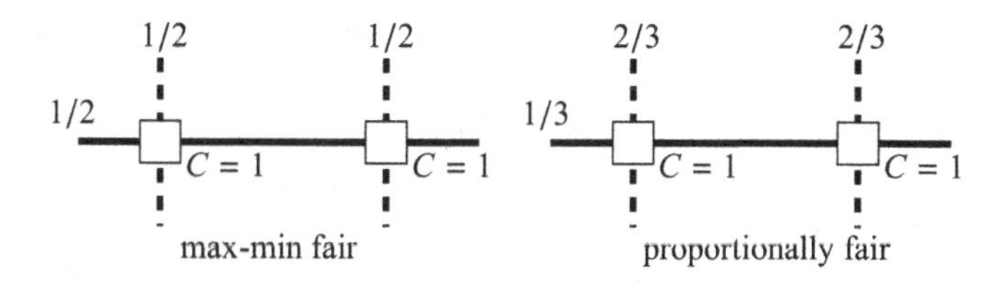

Figure 7.2 The difference between max-min and proportional fairness. Note that the allocation maximizing total throughput is (1,1,0).

In Figure 7.2 we show the max-min and the proportionally fair allocations for a particular network example with two resources, each of unit capacity, and three routes. The throughput-optimal allocation is simply to award all the capacity to the single-resource routes; this will have a throughput of 2. The proportionally fair allocation has a total throughput of 1.67, and the max-min fair one has a total throughput of 1.5.

There is a relationship between the notion of (weighted) proportional fairness and the NETWORK problem we introduced in Section 7.1.

Proposition 7.4 *A vector x solves* NETWORK$(A, C; w)$ *if and only if it is weighted proportionally fair.*

Proof The objective function of NETWORK is strictly concave and continuously differentiable, the feasible region is compact and convex, and the vector x that solves NETWORK is unique. Consider a perturbation of x, $y = x + \delta x$ (so $y_r = x_r + \delta x_r$). The objective function $\sum_{r\in\mathcal{R}} w_r \log x_r$ of NETWORK will change by an amount

$$\sum_{r\in\mathcal{R}} w_r \frac{\delta x_r}{x_r} + o(\delta x).$$

But the first term is simply the expression (7.4). The result follows, using the concavity of the objective function to deduce the inequality (7.4) for non-infinitesimal variations from x. □

Remark 7.5 The proportionally fair allocation has various other related interpretations, including as a solution to Nash's bargaining problem and as a market clearing equilibrium.

Bargaining problem, Nash (1950). Consider the problem of several players bargaining among a set of possible choices. The Nash bargaining solution is the unique vector satisfying the following axioms:

- Invariance under scaling. If we rescale the feasible region, the bargaining solution should simply rescale.
- Pareto efficiency. An allocation is Pareto inefficient if there exists an alternative allocation that improves the amount allocated to at least one player without reducing the amount allocated to any other players.
- Symmetry. If the feasible region is symmetric, then the players should be allocated an equal amount.
- Independence of Irrelevant Alternatives. If we restrict the set of feasible allocations to some subset that contains the original maximum, then the maximum should not change.

A general set of weights w corresponds to a situation with unequal bargaining power; this will modify the notion of symmetry.

Market clearing equilibrium (Gale (1960), Section 8.5). A market-clearing equilibrium is a set of resource prices $p = (p_j, j \in \mathcal{J})$ and an allocation $x = (x_r, r \in \mathcal{R})$ such that

- $p \geq 0, Ax \leq C$ (feasibility);
- $p^T(C - Ax) = 0$ (complementary slackness): if the price of a resource is positive then the resource is used up;
- $w_r = x_r \sum_{j \in r} p_j, r \in \mathcal{R}$: if user r has an endowment w_r, then all endowments are spent.

Exercises

Exercise 7.4 In this exercise, we show that the Nash bargaining solution associated with the problem of assigning flows x_r subject to the constraints $x \geq 0, Ax \leq C$ is the solution of NETWORK$(A, C; w)$ with $w_r = 1$ for all $r \in \mathcal{R}$.

(1) Show that the feasible region can be rescaled by a linear transformation so that the maximum of $\prod_r x_r$ is located at the point $(1, \ldots, 1)$.

(2) Argue that this implies that the reparametrized region must lie entirely below the hyperplane $\sum_r x_r = |\mathcal{R}|$, and hence that $(1, \ldots, 1)$ is a Pareto-efficient point.

(3) By considering a suitable symmetric set that contains the feasible region, and applying Independence of Irrelevant Alternatives, argue that the bargaining solution to the reparametrized problem must be equal to $(1, \ldots, 1)$.

(4) Conclude that the bargaining solution to the original problem must have been the point that maximizes $\prod_r x_r$, or, equivalently, the point that maximizes $\sum_r \log x_r$.

Exercise 7.5 Show that a market-clearing equilibrium (p, x) can be found by solving NETWORK$(A, C; w)$ for x, with p identified as the vector of Lagrange multipliers (or shadow prices) associated with the resource constraints.

7.3 A primal algorithm

We now look at an algorithm that can accomplish fair sharing of resources in a distributed manner. We begin by describing the mechanics of data transfer over the Internet.

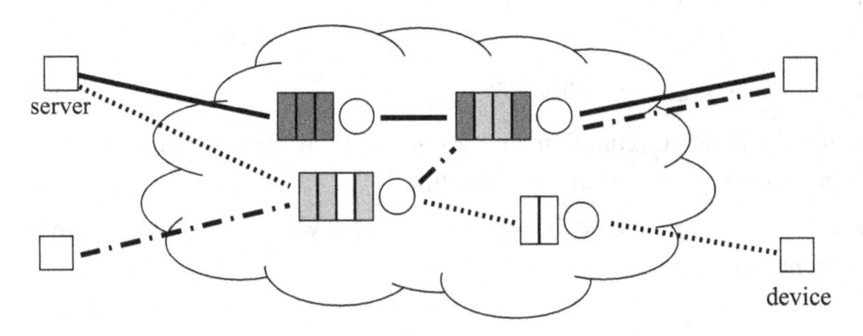

Figure 7.3 A schematic diagram of the Internet. Squares correspond to sources and destinations, the network contains resources with buffers, and many flows traverse the network.

The transfer of a file (for example, a web page) over the Internet begins when the device wanting the file sends a packet with a request to the server that has the file stored. The file is broken up into packets, and the first packet is sent back to the device. If the packet is received successfully, an acknowledgement packet is sent back to the server, and this prompts further packets to be sent. This process is controlled by TCP, the *transmission control protocol* of the Internet. TCP is part of the software of the machines that are the source and destination of the data (the server and the device in our example). The general approach is as follows. When a resource within the network becomes heavily loaded, one or more packets is lost or marked. The lost or marked packet is taken as an indication of congestion, the destination informs the source, and the source slows down. TCP then gradually increases the sending rate until it receives another indication of congestion. This cycle of increase and decrease serves to discover and utilize available capacity, and to share it between flows.

A consequence of this approach is that "intelligence" and control are end-to-end rather than hidden in the network: the machines within the network forward packets, but the machines that are the source and destination regulate the rate. We will now describe a model that attempts to capture this notion of end-to-end control.

Consider a network with parameters A, C, w as before, but now suppose the flow rate $x_r(t)$ on route r is time varying. We define the *primal algorithm* to be the system of differential equations

$$\frac{d}{dt}x_r(t) = \kappa_r \left(w_r - x_r(t) \sum_{j \in r} \mu_j(t) \right), \qquad r \in \mathcal{R}, \tag{7.5}$$

where

$$\mu_j(t) = p_j \left(\sum_{s: j \in s} x_s(t) \right), \qquad j \in \mathcal{J}, \tag{7.6}$$

$w_r, \kappa_r > 0$ for $r \in \mathcal{R}$ and the function $p_j(\cdot)$ is a non-negative, continuous and increasing function, not identically zero, for $j \in \mathcal{J}$.

The primal algorithm is a useful caricature of end-to-end control: congestion indication signals are generated at resource j at rate $\mu_j(t)$; these signals reach each user r whose route passes through resource j; and the response to this feedback by the user r is a decrease at a rate proportional to the stream of feedback signals received together with a steady increase at rate proportional to the weight w_r. We could imagine that the function $p_j(\cdot)$ measures the "pain" of resource j as a function of the total traffic going through it. (For example, it could be the probability of a packet being dropped or marked at resource j.)

Note the local nature of the primal algorithm: the summation in equation (7.5) concerns only resources j that are used by route r, and the summation in equation (7.6) concerns only routes s that pass through resource j. Nowhere in the network is there a need to know the entire link-route incidence matrix A.

Starting from an initial state $x(0)$, the dynamical system (7.5)–(7.6) defines a trajectory $(x(t), t \geq 0)$. We next show that, whatever the initial state, the trajectory converges to a limit, and that the limit solves a certain optimization problem. We shall establish this by exhibiting a *Lyapunov function* for the system of differential equations. (A Lyapunov function is a scalar function of the state of the system, $x(t)$, that is monotone in time t along trajectories.)

Theorem 7.6 (Global stability) *The strictly concave function*

$$\mathcal{U}(x) = \sum_{r \in \mathcal{R}} w_r \log x_r - \sum_{j \in \mathcal{J}} \int_0^{\sum_{s: j \in s} x_s} p_j(y) dy$$

is a Lyapunov function for the primal algorithm. The unique value maximizing $\mathcal{U}(x)$ is an equilibrium point of the system, to which all trajectories converge.

Proof The assumptions on $w_r > 0, r \in \mathcal{R}$, and $p_j(\cdot), j \in \mathcal{J}$, ensure that $\mathcal{U}(\cdot)$ is strictly concave with a maximum that is interior to the positive orthant; the maximizing value of x is thus unique. Moreover, $\mathcal{U}(x)$ is continuously differentiable with

$$\frac{\partial}{\partial x_r}\mathcal{U}(x) = \frac{w_r}{x_r} - \sum_{j\in r} p_j\left(\sum_{s:j\in s} x_s\right),$$

and setting these derivatives to zero identifies the maximum, $\overline{x}$ say. The derivative (7.5) is zero at $\overline{x}$, and hence $\overline{x}$ is an equilibrium point.

Further,

$$\begin{aligned}\frac{d}{dt}\mathcal{U}(x(t)) &= \sum_{r\in\mathcal{R}} \frac{\partial \mathcal{U}}{\partial x_r}\frac{d}{dt}x_r(t) \\ &= \sum_{r\in\mathcal{R}} \frac{\kappa_r}{x_r(t)}\left(w_r - x_r(t)\sum_{j\in r} p_j\left(\sum_{s:j\in s} x_s(t)\right)\right)^2 \geq 0,\end{aligned} \tag{7.7}$$

with equality only at $\overline{x}$. This almost, but not quite, proves the desired convergence to $\overline{x}$: we need to make sure that the derivative (7.7) isn't so small that the system grinds to a halt far away from $\overline{x}$. However, this is easy to show.

Consider an initial state $x(0)$ in the interior of the positive orthant. The trajectory $(x(t), t \geq 0)$ cannot leave the compact set $\{x : \mathcal{U}(x) \geq \mathcal{U}(x(0))\}$. Consider the complement, C, within this compact set of an open ϵ-ball centred on $\overline{x}$. Then C is compact, and so the continuous derivative (7.7) is bounded away from zero on C. Hence the trajectory can only spend a finite time in C before entering the ϵ-ball. Thus, $x(t) \to \overline{x}$. □

We have shown that the primal algorithm optimizes the function $\mathcal{U}(x)$. Now we can view

$$C_j\left(\sum_{s:j\in s} x_s\right) = \int_0^{\sum_{s:j\in s} x_s} p_j(y)dy$$

as a cost function penalizing use of resource j. If we take

$$p_j(y) = \begin{cases}\infty, & y > C_j, \\ 0, & y \leq C_j,\end{cases}$$

then maximizing the Lyapunov function $\mathcal{U}(x)$ becomes precisely the problem NETWORK$(A, C; w)$. While this choice of p_j violates the continuity assumption on p_j, we could approximate it arbitrarily closely by smooth functions, and so it would seem that this approach will generate arbitrarily good approximations of the NETWORK problem. However, we shall see later that, in the presence of propagation delays in the network, large values of $p'_j(\cdot)$ compromise stability, so there is a limit to how closely we can approximate the NETWORK problem using this approach.

Exercises

Exercise 7.6 Suppose that the parameter κ_r in the primal algorithm is replaced by a function $\kappa_r(x_r(t))$, for $r \in \mathcal{R}$. Check that Theorem 7.6 remains valid, with an unaltered proof, for any initial state $x(0)$ in the interior of the positive orthant provided the functions $\kappa_r(\cdot), r \in \mathcal{R}$, are continuous and positive on the interior of the positive orthant. (In Section 7.7 we shall be interested in the choice $\kappa_r(x_r(t)) = \kappa_r x_r(t)/w_r$.)

Exercise 7.7 Suppose there really *is* a cost $C_j(z) = z/(C_j - z)$ to carrying a load z through resource j, and that it is possible to choose the functions $p_j(\cdot)$. (Assume $C_j(z) = \infty$ if $z \geq C_j$.) Show that the sum of user utilities minus costs is maximized by the primal algorithm with the choices $p_j(z) = C_j/(C_j - z)^2, j \in \mathcal{J}$.

The above function $C_j(\cdot)$ arises from a simple queueing model of a resource (see Section 4.3.1). Next we consider a simple time-slotted model of a resource.

Suppose we model the number of packets arriving at resource j in a time slot as a Poisson random variable with mean z, and that if the number is above a limit N_j there is some cost to dealing with the excess, so that

$$C_j(z) = e^{-z} \sum_{n>N_j} (n - N_j) \frac{z^n}{n!}.$$

Show that the sum of user utilities minus costs is maximized by the primal algorithm with the choices

$$p_j(z) = e^{-z} \sum_{n \geq N_j} \frac{z^n}{n!}.$$

Observe that p_j will be the proportion of packets marked at resource j if the following mechanism is adopted: mark *every* packet in any time slot in which N_j or more packets arrive.

Exercise 7.8 In this exercise, we generalize the primal algorithm to the case where $s \in S$ labels a source–destination pair served by routes $r \in s$, as in Exercise 7.2.

Let $s(r)$ identify the unique source–destination pair served by route r. Suppose that the flow along route r evolves according to

$$\frac{d}{dt}y_r(t) = \kappa_r \left(w_{s(r)} - \left(\sum_{a \in s(r)} y_a(t) \right) \sum_{j \in r} \mu_j(t) \right)$$

(or zero if this expression is negative and $y_r(t) = 0$), where

$$\mu_j(t) = p_j \left(\sum_{r: j \in r} y_r(t) \right),$$

and let

$$\mathcal{U}(y) = \sum_{s \in S} w_s \log \left(\sum_{r \in s} y_r \right) - \sum_{j \in \mathcal{J}} \int_0^{\sum_{r:j \in r} y_r} p_j(z) dz.$$

Show that

$$\frac{d}{dt}\mathcal{U}(y(t)) > 0$$

unless y solves the problem NETWORK$(H, A, C; w)$ from Exercise 7.2.

7.4 Modelling TCP

Next we describe in slightly more detail the congestion avoidance algorithm of TCP, due to Jacobson (1988). Let T, the *round-trip time*, be the time between a source sending a packet and the source receiving an acknowledgement. The source attempts to maintain a window (of size cwnd) of packets that have been sent but not yet acknowledged. The rate x of our model represents the ratio cwnd$/T$. If the acknowledgement is positive, cwnd is increased by $1/$cwnd, while if the acknowledgement is negative (a packet was lost or marked), cwnd is halved.

Remark 7.7 Even our more detailed description of TCP is simplified, omitting discussion of timeouts (which trigger retransmissions, with binary exponential backoff) and the slow start phase (during which the sending rate grows exponentially). But the above description is sufficient for us to develop a system of differential equations to compare with those of Section 7.3.

Figure 7.4 shows the typical evolution of window size, in increments of the round-trip time, for TCP. Modelling this behaviour by a differential equation is at first sight implausible: the rate x is very clearly *not* smooth. It is helpful to think first of a weighted version of TCP, MulTCP, due

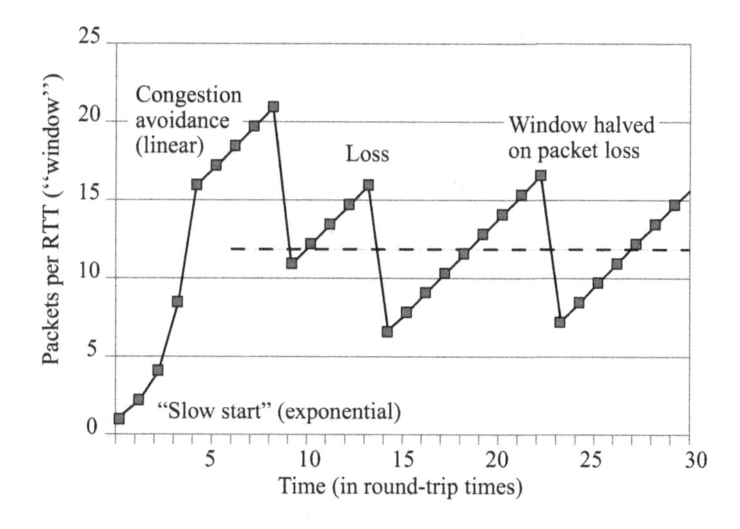

Figure 7.4 Typical window size for TCP.

to Crowcroft and Oechslin (1998), which is in general smoother. Let w be a weight parameter, and suppose:

- the rate of additive increase is multiplied by w, so that each acknowledgement increases $\mathtt{cwnd}$ by $w/\mathtt{cwnd}$; and
- the multiplicative decrease factor becomes $1 - 1/(2w)$, so that after a congestion indication the window size becomes $(1 - 1/(2w))\mathtt{cwnd}$.

The weight parameter w is designed to imitate the user sending traffic via w distinct TCP streams; the original algorithm corresponds to $w = 1$. This modification results in a smoother trajectory for the rate x for larger values of w. It is also a crude model for the aggregate of w distinct TCP flows over the same route. (Observe that if a congestion indication is received by one of w distinct TCP flows then only one flow will halve its window.)

Let us try to approximate the rate obtained by MulTCP by a differential equation. Let p be the probability of a congestion indication being received during an update step. The expected change in the congestion window $\mathtt{cwnd}$ per update step is approximately

$$\frac{w}{\mathtt{cwnd}}(1 - p) - \frac{\mathtt{cwnd}}{2w}p. \tag{7.8}$$

Since the time between update steps is approximately $T/\mathtt{cwnd}$, the expected change in the rate x per unit time is approximately

$$\frac{\left(\frac{w}{\texttt{cwnd}}(1-p) - \frac{\texttt{cwnd}}{2w}p\right)/T}{T/\texttt{cwnd}} = \frac{w}{T^2} - \left(\frac{w}{T^2} + \frac{x^2}{2w}\right)p.$$

Motivated by this calculation, we model MulTCP by the system of differential equations

$$\frac{d}{dt}x_r(t) = \frac{w_r}{T_r^2} - \left(\frac{w_r}{T_r^2} + \frac{x_r^2}{2w_r}\right)p_r(t),$$

where

$$p_r(t) = \sum_{j \in r} \mu_j(t)$$

and $\mu_j(t)$ is given by (7.6). Here, T_r is the round-trip time and w_r is the weight parameter for route r. If congestion indication is provided by dropping a packet then $p_r(t)$ approximates the probability of a packet drop along a route by the sum of the packet drop probabilities at each of the resources along the route.

Theorem 7.8 (Global stability)

$$\mathcal{U}(x) = \sum_{r \in \mathcal{R}} \frac{\sqrt{2}w_r}{T_r} \arctan\left(\frac{x_r T_r}{\sqrt{2}w_r}\right) - \sum_{j \in \mathcal{J}} \int_0^{\sum_{s:j \in s} x_s} p_j(y)dy$$

is a Lyapunov function for the above system of differential equations. The unique value x maximizing $\mathcal{U}(x)$ is an equilibrium point of the system, to which all trajectories converge.

Proof Observe that

$$\frac{\partial}{\partial x_r}\mathcal{U}(x) = \frac{w_r}{T_r^2}\left(\frac{w_r}{T_r^2} + \frac{x_r^2}{2w_r}\right)^{-1} - \sum_{j \in r} p_j\left(\sum_{s:j \in s} x_s\right)$$

and so

$$\frac{d}{dt}\mathcal{U}(x(t)) = \sum_{r \in R} \frac{\partial \mathcal{U}}{\partial x_r}\frac{d}{dt}x_r(t) = \sum_{r \in R}\left(\frac{w_r}{T_r^2} + \frac{x_r^2}{2w_r}\right)^{-1}\left(\frac{d}{dt}x_r(t)\right)^2 \geq 0.$$

The proof proceeds as before. □

At the stable point,

$$x_r = \frac{w_r}{T_r}\left(2\frac{1-p_r}{p_r}\right)^{1/2}.$$

If we think of x_r as the aggregate of w_r distinct TCP flows on the same route, then each of these flows has an average throughput that is given by this expression with $w_r = 1$.

The constant $2^{1/2}$ appearing in this expression is sensitive to the difference between MulTCP and w distinct TCP flows. With w distinct TCP flows, the flow affected by a congestion indication is more likely to be one with a larger window. This bias towards the larger of the w windows increases the final term of the expectation (7.8) and decreases the constant, but to prove this would require a more sophisticated analysis, such as in Ott (2006).

Remark 7.9 It is instructive to compare our modelling approach in this chapter with that of earlier chapters. Our models of a loss network in Chapter 3, and of an electrical network in Chapter 4, began with detailed probabilistic descriptions of call acceptance and electron motion, respectively, and we derived macroscopic relationships for blocking probabilities and currents such as the Erlang fixed point and Ohm's law. In this chapter we have not developed a detailed probabilistic description of queueing behaviour at resources within the network: instead we have crudely summarized this behaviour with the functions $p_j(\cdot)$. We have adopted this approach since the control exerted by end-systems, the additive increase and multiplicative decrease rules for window sizes, is the overwhelming influence at the packet level on flow rates. It is these rules that lead to the particular form of a flow rate's dependence on T_r and p_r, which we'll explore in more detail in the next section.

Exercise

Exercise 7.9 Check that if p_r is small, corresponding to a low end-to-end loss probability, then at the stable point

$$x_r = \frac{w_r}{T_r}\sqrt{\frac{2}{p_r}} + o(p_r).$$

Check that if x_r is large, again corresponding to a low end-to-end loss probability, then

$$U_r(x_r) = \frac{\sqrt{2}w_r}{T_r}\arctan\left(\frac{x_r T_r}{\sqrt{2}w_r}\right) = \text{const.} - \frac{2w_r^2}{T_r^2 x_r} + o\left(\frac{1}{x_r}\right).$$

7.5 What is being optimized?

From Theorem 7.4 we see that TCP (or at least our differential equation model of it) is behaving as if it is maximizing the sum of user utilities, subject to a cost function penalizing proximity to the capacity constraints. The user utility for a single TCP flow ($w_r = 1$) is, from Exercise 7.9, of the form

$$U_r(x_r) \approx \text{const.} - \frac{2}{T_r^2 x_r}.$$

Further, the rate allocated to a single flow on route r is approximately proportional to

$$\frac{1}{T_r p_r^{1/2}}, \tag{7.9}$$

so that the rate achieved is inversely proportional to the round-trip time T_r and to the square root of the packet loss probability p_r.

The proportionally fair allocation of rates would be inversely proportional to p_r and would have no dependence on T_r. Relative to proportional fairness, the above rate penalizes T_r and under-penalizes p_r: it penalizes distance and under-penalizes congestion.

For many files downloaded over the Internet, the round-trip times are negligible compared to the time it takes to download the file: the round-trip time is essentially a speed-of-light delay, and is of the order of 50 ms for a transatlantic connection, but may be much smaller. If limited capacity is causing delay in downloading a file, it would be reasonable to expect that the user would not care very much whether the round-trip time is in fact 1 ms or 50 ms – and we would not expect a big difference in user utility.

Another way to illustrate the relative impact of distance and congestion is the problem of *cache location*. Consider the network in Figure 7.5. On the short route, from a user to cache location A, the round-trip time is $T = 2T_1$, and the probability of congestion indication is $p \approx 2p_1$ (we are assuming p_1 and p_2 are small). On the long route, from the user to cache location B, we have $T = T_1 + T_2$ and $p \approx p_1 + p_2$.

Suppose now that link 2 is a high-capacity long-distance optical fibre. Plausible parameters for it are $p_2 = 0$ (resource 2 is underused), but $T_2 = 100T_1$. Then the ratio of TCP throughputs along the two routes is

$$\frac{T_1 + T_2}{2T_1} \frac{\sqrt{p_1 + p_2}}{\sqrt{2p_1}} = \frac{101}{2\sqrt{2}} \approx 36.$$

That is, the throughput along the short route is much higher than that along

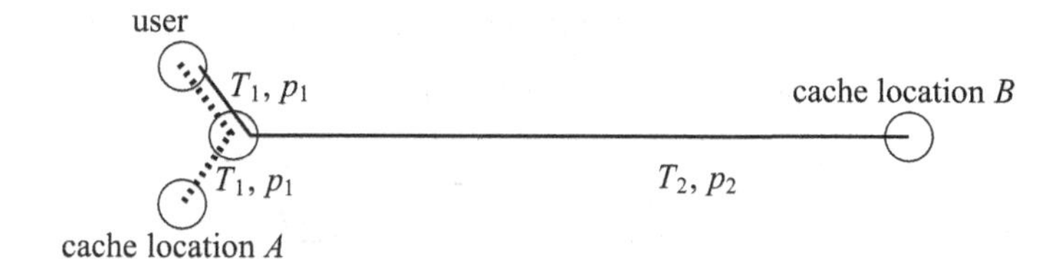

Figure 7.5 A network with three links and two routes: long (solid) and short (dotted). The round-trip times and packet drop probabilities are indicated on the links.

the long one, even though the short route is using two congested resourcesand the long route uses only one.

Suppose that we want to place a cache with lots of storage space somewhere in the network, and must choose between putting it at A or at B. The cache will appear more effective if we put it at A, further loading the already congested resources. This is an interesting insight into consequences of the implicit optimization inherent in the form (7.9) – the dependence of throughput on round-trip times means that decentralized decisions on caching will tend to overload the congested edges of the network and underload the high-capacity long-distance core.

Exercises

Exercise 7.10 What is the ratio of throughputs along the two routes in Figure 7.5 under proportional fairness?

Exercise 7.11 Our models of congestion control assume file sizes are large, so that there is time to control the flow rate while the file is in the process of being transferred. Many files transferred by TCP have only one or a few packets, and are essentially uncontrolled (*mice* rather than *elephants*). Suppose the uncontrolled traffic through resource j is u_j, and $p_j(y)$ is replaced by $p_j(y + u_j)$ in our systems of differential equations. Find the function $\mathcal{U}$ that is being implicitly optimized.

(*Note:* There will be good reasons for placing mice, if not elephants, in a cache within a short round-trip time of users.)

7.6 A dual algorithm

Recall the optimization problem $\text{Dual}(A, C; w)$:

$$\begin{aligned} &\text{maximize} \quad \sum_{r\in\mathcal{R}} w_r \log\left(\sum_{j\in r} \mu_j\right) - \sum_{j\in J} \mu_j C_j \\ &\text{over} \quad \mu \geq 0. \end{aligned}$$

Define the *dual algorithm*

$$\frac{d}{dt}\mu_j(t) = \kappa_j \mu_j(t) \left(\sum_{r:j\in r} x_r(t) - C_j\right), \qquad j \in \mathcal{J},$$

where

$$x_r(t) = \frac{w_r}{\sum_{k\in r} \mu_k(t)}, \qquad r \in \mathcal{R},$$

and $\kappa_j > 0, j \in \mathcal{J}$. That is, the rate of change of μ_j is proportional to μ_j and to the *excess demand* for link j. In economics, such a process is known as a "tâtonnement process" ("tâtonnement" is French for "groping"). It is a natural model for a process that attempts to balance supply and demand.

For the dual algorithm, the intelligence of the system is at the resources $j \in \mathcal{J}$, rather than at the users $r \in \mathcal{R}$. That is, the end systems simply maintain $x_r(t) = 1/\sum_{k\in r} \mu_k(t)$ and it is the resources that have the trickier task of adjusting $\mu_j(t)$. In the primal algorithm, by contrast, the intelligence was placed with the end systems.

Let us check that the dual algorithm solves the Dual problem. We do this by showing the objective function of Dual is a Lyapunov function. Let

$$\mathcal{V}(\mu) = \sum_{r\in\mathcal{R}} w_r \log\left(\sum_{j\in r} \mu_j\right) - \sum_{j\in\mathcal{J}} \mu_j C_j,$$

and let $\mu(0)$ be an initial state in the interior of the positive orthant. Then

$$\frac{d}{dt}\mathcal{V}(\mu(t)) = \sum_{j\in\mathcal{J}} \frac{\partial \mathcal{V}}{\partial \mu_j} \frac{d}{dt}\mu_j(t) = \sum_{j\in\mathcal{J}} \kappa_j \mu_j(t) \left(\sum_{r:j\in r} x_r(t) - C_j\right)^2 \geq 0,$$

with equality only at a point which is both an equilibrium point of the dual algorithm and a maximum of the concave function $\mathcal{V}(\mu)$. If the matrix A is of full rank then the function $\mathcal{V}(\mu)$ is strictly concave and there is a unique optimum, to which trajectories necessarily converge.

Exercises

Exercise 7.12 Suppose that the matrix A is not of full rank. Show that, whereas the value of μ maximizing the concave function $\mathcal{V}(\mu)$ may not be unique, the vector $\mu^T A$ *is* unique, and hence the optimal flow pattern $\bar{x}$ is unique. Prove that, along any trajectory $(\mu(t), t \geq 0)$ of the dual algorithm starting from an initial state $\mu(0)$ in the interior of the positive orthant, $x(t)$ converges to $\bar{x}$.
[*Hint:* Consider the set of points μ maximizing the function $\mathcal{V}(\mu)$, and show that $\mu(t)$ approaches this set.]

Exercise 7.13 Consider a generalized model of the dual algorithm, where the flow rate is given by

$$x_r(t) = D_r\left(\sum_{j \in r} \mu_j(t)\right).$$

The function $D_r(\cdot)$ can be interpreted as the demand for traffic on route r: assume it is a positive, strictly decreasing function of the total price on route r.

Let

$$\mathcal{V}(\mu) = \sum_{r \in \mathcal{R}} \int^{\sum_{j \in r} \mu_j} D_r(\xi) d\xi - \sum_{j \in \mathcal{J}} \mu_j C_j.$$

By considering $\partial \mathcal{V} / \partial \mu_j$, show that $d\mathcal{V}(\mu(t))/dt$ will be non-negative provided we set

$$\frac{d}{dt}\mu_j(t) \gtrless 0 \text{ according as } \sum_{r: j \in r} x_r(t) \gtrless C_j.$$

This suggests we can choose from a large family of price update mechanisms, provided we increase the price of those resources where the capacity constraint is binding, and decrease the price of those resources where it is not. We shall see that in the presence of time delays this freedom is actually somewhat limited.

7.7 Time delays

In earlier sections, the parameters $\kappa_r, r \in \mathcal{R}$, or $\kappa_j, j \in \mathcal{J}$, were assumed to be positive, but were otherwise arbitrary. This freedom was possible since the feedback of information between resources and sources was assumed to be instantaneous. In this section, we briefly explore what happens when feedback is assumed to be fast, e.g. at the speed of light, but nonetheless finite.

Consider first a very simple, single-resource model of the primal algorithm, but one where the change in flow rate at time t depends on congestion information from a time T units earlier.

$$\frac{d}{dt}x(t) = \kappa\Big(w - x(t-T)p(x(t-T))\Big).$$

In the Internet, a natural value for T is the round-trip time; information about the fate of packets on route r takes approximately that long to reach the source.

This system has an equilibrium point $x(t) = x_e$, which solves

$$w = x_e p(x_e).$$

Our interest is in how the system behaves if we introduce a small fluctuation about this rate. Write $x(t) = x_e + u(t)$. Let $p_e = p(x_e)$ and $p'_e = p'(x_e)$. Linearizing the time-delayed equation about x_e, by neglecting second-order terms in $u(\cdot)$, we obtain

$$\begin{aligned}\frac{d}{dt}u(t) &\approx \kappa\Big(w - (x_e + u(t-T))(p_e + p'_e u(t-T))\Big)\\ &\approx -\kappa(p_e + x_e p'_e)u(t-T).\end{aligned}$$

Now, the solutions to a single-variable linear differential equation with delay,

$$\frac{d}{dt}u(t) = -\alpha u(t-T), \tag{7.10}$$

are quite easy to find. Figure 7.6 plots the qualitative behaviour of the solutions, as the *gain* α increases from 0 to ∞.

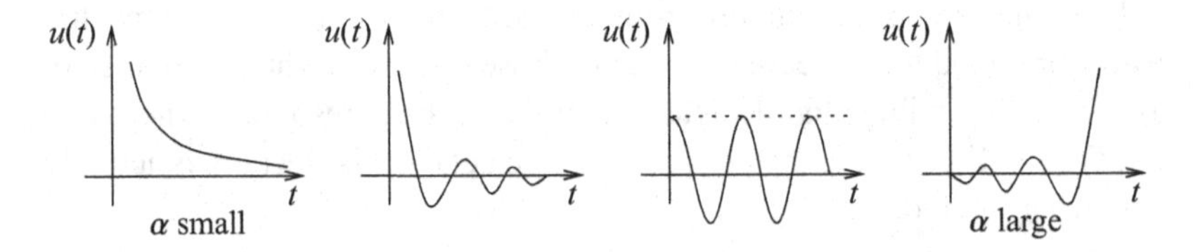

Figure 7.6 Solutions of the differential equation for different values of α.

To find the value α_c that corresponds to the periodic solution, try using $u(t) = \sin(\lambda t)$. Then

$$\lambda \cos \lambda t = -\alpha_c \sin \lambda(t-T) = -\alpha_c(\sin \lambda t \cos \lambda T - \cos \lambda t \sin \lambda T),$$

which gives $\cos \lambda T = 0$ and

$$\alpha_c = \lambda = \frac{\pi}{2}\frac{1}{T}.$$

The periodic solution marks the boundary between stability and instability of equation (7.10), and hence *local* stability and instability of the earlier non-linear system.

Remark 7.10 Note that we could have concluded $\alpha_c \propto 1/T$ from dimensional analysis alone: given that the critical value for α exists, it has dimension of inverse time, and T is the only time parameter in the problem. For our single-resource model, the condition for local stability about the equilibrium point becomes

$$\kappa T(p_e + x_e p_e') < \frac{\pi}{2}.$$

Thus a larger derivative p_e' requires a smaller gain κ.

The argument suggests that the gain κ in the update mechanism should be limited by the fact that the product κT should not get too large. Of course, in the absence of time delays, a larger value for κ corresponds to faster convergence to equilibrium, so we would like κ to be as large as possible.

We briefly comment, without proofs, on the extension of these ideas to a network. For each j, r such that $j \in r$, let T_{rj} be the time delay for packets from the source for route r to reach the resource j, and let T_{jr} be the return delay for information from the resource j to reach the source. We suppose these delays satisfy

$$T_{rj} + T_{jr} = T_r, \qquad j \in r,\ r \in \mathcal{R},$$

where T_r is the round-trip time on route r.

Consider the time-delayed dynamical system

$$\frac{d}{dt}x_r(t) = \kappa_r x_r(t - T_r)\left(1 - p_r(t) - p_r(t)\frac{x_r(t)}{w_r}\right) \tag{7.11}$$

where

$$p_r(t) = 1 - \prod_{j \in r}\left(1 - \mu_j(t - T_{jr})\right)$$

and

$$\mu_j(t) = p_j\left(\sum_{s: j \in s} x_s(t - T_{sj})\right),$$

and $\kappa_r > 0, r \in \mathcal{R}$. (This model arises in connection with variants of TCP: comparing the right-hand side of equation (7.11) with expression (7.8), we see that $x_r(t - T_r)(1 - p_r(t))$ and $x_r(t - T_r)p_r(t)$ represent the stream of returning positive and negative acknowledgements.) If packets are dropped or marked at resource j with probability p_j independently from other resources, then p_r measures the probability that a packet traverses the entire route r intact. To see where the delays enter, note that the feedback seen at the source for route r at time t is carried on a flow that passed through the resource j a time T_{jr} previously, and the flow rate on route s that is seen at resource j at time t left the source for route s a time T_{sj} previously.

A plausible choice for the marking probability $p_j(\cdot)$ is

$$p_j(y_j) = \left(\frac{y_j}{C_j}\right)^B;$$

this is the probability of finding a queue of size $\geq B$ in an $M/M/1$ queue with arrival rate y_j and service rate C_j. (While we don't expect the individual resources to behave precisely like independent $M/M/1$ queues with fixed arrival rates, this is a sensible heuristic.) It can be shown that the system of differential equations with delay will be locally stable about its equilibrium point provided

$$\kappa_r T_r B < \frac{\pi}{2}, \qquad \forall r \in \mathcal{R}.$$

This is a strikingly simple result: the gain parameter for route r is limited only by the round-trip time T_r, and not by the round-trip times of other routes. As B increases, and thus as the function $p_j(\cdot)$ approaches a sharp transition at the capacity C_j, the gain parameters must decrease.

For the dual algorithm of Section 7.6, the time-delayed dynamical system becomes

$$\frac{d}{dt}\mu_j(t) = \kappa_j \mu_j(t)\left(\sum_{r:j\in r} x_r(t - T_{rj}) - C_j\right),$$

where

$$x_r(t) = \frac{w_r}{\sum_{k\in r} \mu_k(t - T_{kr})}$$

and $\kappa_j > 0, j \in \mathcal{J}$. It can be shown that this system will be locally stable about its equilibrium point provided

$$\kappa_j C_j \overline{T}_j < \frac{\pi}{2}, \qquad \forall j \in \mathcal{J}, \tag{7.12}$$

where

$$\overline{T}_j = \frac{\sum_{r:j\in r} x_r T_r}{\sum_{r:j\in r} x_r},$$

the average round-trip time of packets through resource j. Again we see a strikingly simple result, where the gain parameter at resource j is limited by characteristics local to resource j.

Exercises

Exercise 7.14 In the single-resource model of delay (7.10), find the value of α that corresponds to the transition from the first to the second graph in Figure 7.6.
[*Hint:* Look for solutions of the form $u(t) = e^{-\lambda t}$ with λ real. *Answer*: $\alpha = 1/(eT)$.]

Exercise 7.15 Check that at an equilibrium point for the system (7.11)

$$x_r = w_r \frac{1 - p_r}{p_r},$$

and is thus independent of T_r.

Exercise 7.16 The simple, single-resource model of the dual algorithm becomes

$$\frac{d}{dt}\mu(t) = \kappa\mu(t)\left(\frac{w}{\mu(t-T)} - C\right).$$

Linearize this equation about its equilibrium point, to obtain an equation of the form (7.10): what is α? Check that the condition $\alpha T < \pi/2$ for local stability is a special case of equation (7.12).

7.8 Modelling a switch

In this chapter we have described packet-level algorithms implemented at end-points, for example in our discussion of TCP. Our model of resources within the network has been very simple, captured by linear constraints of the form $Ax \leq C$. In this section, we shall look briefly at a more developed model of an Internet router, and see how these linear constraints could emerge.

Consider then the following model. There is a set of queues indexed by $\mathcal{R}$. Packets arrive into queues as independent Poisson processes with rates λ_r. There are certain constraints on the sets of queues that can be served simultaneously. Assume that at each discrete time slot there is a finite set $\mathcal{S}$ of possible schedules, only one of which can be chosen. A schedule σ is a vector of non-negative integers $(\sigma_r)_{r\in\mathcal{R}}$, which describes the number of packets to be served from each queue. We assume that the schedule provides an upper bound; that is, if σ is a schedule, and $\pi \leq \sigma$ componentwise, then π is also a schedule.

We are interested in when this system may be stable. Write $\lambda \in \Lambda$ if λ is non-negative and if there exist constants c_σ, $\sigma \in \mathcal{S}$, with $c_\sigma \geq 0$, $\sum c_\sigma = 1$, such that $\lambda \leq \sum c_\sigma \sigma$. Call Λ the *admissible region.*

We would not expect the system to be stable if the arrival rate vector λ lies outside of Λ, since then there is a weighted combination of workloads in the different queues whose drift is upwards whatever scheduling strategy is used (Exercise 7.17).

Example 7.11 (An input-queued switch) An Internet router has N input ports and N output ports. A data transmission cable is attached to each of these ports. Packets arrive at the input ports. The function of the router is to work out which output port each packet should go to, and to transfer packets to the correct output ports. This last function is called switching. There are a number of possible switch architectures; in this example, we will describe the common input-queued switch architecture.

A queue r is held for each pair of an input port and an output port. In each time slot, the switch fabric can transmit a number of packets from input ports to output ports, subject to the constraints that each input can transmit at most one packet and that each output can receive at most one packet. In other words, at each time slot the switch can choose a *matching* from inputs to outputs. Therefore, the set $\mathcal{S}$ of possible schedules is the set of all possible matchings between inputs and outputs (Figure 7.7). The indexing set $\mathcal{R}$ is the set of all input–output pairs.

A matching identifies a permutation of the set $\{1, 2, \ldots, N\}$, and so for this example we can identify $\mathcal{S}$ with the set of permutation matrices.The Birkhoff–von Neumann theorem states that the convex hull of $\mathcal{S}$ is the set of doubly stochastic matrices (i.e. matrices with non-negative entries such that all row and column sums are 1). We deduce that for an input-queued switch

$$\Lambda = \Big\{\lambda \in [0,1]^{NxN} \quad : \quad \sum\nolimits_{i=1}^{N} \lambda_{ij} \leq 1, \quad j = 1, 2, \ldots, N,$$
$$\sum\nolimits_{j=1}^{N} \lambda_{ij} \leq 1, \quad i = 1, 2, \ldots, N\Big\}. \qquad (7.13)$$

Example 7.12 (A wireless network) K stations share a wireless medium:

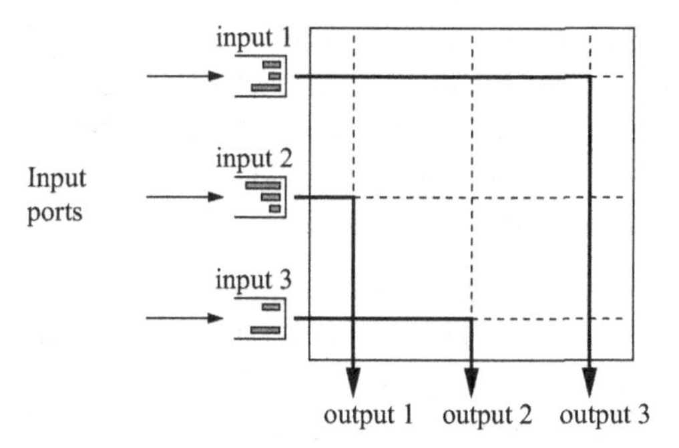

Figure 7.7 Input-queued switch with three input and three output ports, and thus nine queues. One possible matching is highlighted in bold. The figure is from Shah and Wischik (2012), with permission.

the stations form the vertex set of an interference graph, and there is an edge between two stations if they interfere with each other's transmissions. A schedule $\sigma \in \mathcal{S} \subset \{0,1\}^K$ is a vector $\sigma = (\sigma_1, \sigma_2, \dots, \sigma_K)$ with the property that if i and j have an edge between them then $\sigma_i \cdot \sigma_j = 0$.

We've looked at similar networks previously, in Example 2.24 and Exercise 5.2, without explicit coordination between nodes. In this section, we suppose there is some centralized form of scheduling, so that σ can be chosen as a function of queue sizes at the stations.

We return now to the more general model, with queues indexed by $\mathcal{R}$ and schedules indexed by $\mathcal{S}$. We look at a family of algorithms, the MaxWeight-α family, which we shall show stabilizes the system for all arrival rates in the interior of Λ.

Fix $\alpha \in (0, \infty)$. At time t, pick a schedule $\sigma \in \mathcal{S}$ to solve

$$\text{maximize} \sum_{r \in \mathcal{R}} \sigma_r q_r(t)^\alpha$$
$$\text{subject to } \sigma \in \mathcal{S},$$

where $q_r(t)$ is the number of packets in queue r at time t. If the maximizing σ is not unique, pick a schedule at random from amongst the set of schedules that attain the maximum. Note the myopic nature of the algorithm: the schedule chosen at time t depends only on the current queue size vector $q(t) = (q_r(t), r \in \mathcal{R})$, which is thus a Markov process under the assumption of Poisson arrivals.

As $\alpha \to 0$, the strategy aims to maximize the number of packets served, corresponding to throughput maximization. If $\alpha = 1$, the schedule σ maximizes a weighted sum of the number of packets served, with weights the queue lengths (a *maximum weight schedule*). When $\alpha \to \infty$, the schedule σ aims to maximize service to the longest queue, then to the second longest, and so on, reminiscent of max-min fairness.

Theorem 7.13 *For any $\alpha \in (0, \infty)$, and Poisson arrivals with rates $\lambda = (\lambda_r, r \in \mathcal{R})$ that lie in the interior of Λ, the MaxWeight-α algorithm stabilizes the system. That is, the Markov chain $q(t) = (q_r(t), r \in \mathcal{R})$ describing the evolution of queue sizes is positive recurrent.*

Sketch of proof Consider the *Lyapunov function*

$$\mathcal{L}(q) = \sum \frac{q_r^{\alpha+1}}{\alpha + 1}.$$

This is a non-negative function. We argue that the drift of $\mathcal{L}$ should be negative provided $||q(t)||$ is large:

$$\mathbb{E}[\mathcal{L}(q(t+1)) - \mathcal{L}(q(t)) \,|\, q(t) \text{ large}] < 0.$$

(This suggests that $\mathcal{L}(q(t))$ should stay fairly small, and hence that $q(t)$ should be positive recurrent. However, as we have seen in Section 5.2, simply having negative drift is insufficient to conclude positive recurrence of a Markov chain, so to turn the sketch into a proof we need to do more work, later.)

Let us approximate the drift of $\mathcal{L}$ by a quantity that is easier to estimate. When $q(t)$ is large, we can write

$$\mathcal{L}(q(t+1)) \approx \mathcal{L}(q(t)) + \sum_r \frac{\partial \mathcal{L}(q(t))}{\partial q_r} (q_r(t+1) - q_r(t)).$$

Therefore

$$\mathbb{E}[\mathcal{L}(q(t+1)) - \mathcal{L}(q(t)) \,|\, q(t)] \approx \sum_r q_r(t)^\alpha \mathbb{E}[q_r(t+1) - q_r(t) \,|\, q(t)], \quad (7.14)$$

provided $q(t)$ is large.

Let $\sigma^*(t)$ be the MaxWeight-α schedule selected at time t. Then

$$q_r(t)^\alpha \mathbb{E}[q_r(t+1) - q_r(t) \,|\, q(t)] = q_r(t)^\alpha (\lambda_r - \sigma_r^*(t)).$$

(If $q_r(t) = 0$ then $\mathbb{E}[q_r(t+1) - q_r(t) \,|\, q(t)]$ may not be $\lambda_r - \sigma_r^*(t)$, but the product is still zero.)

Now, the definition of $\sigma^*(t)$ means that

$$\sum_r \sigma_r^*(t) q_r(t)^\alpha \geq \sum_r \sum_\sigma c_\sigma \sigma_r q_r(t)^\alpha$$

for any other convex combination of feasible schedules. Since we know there is a convex combination of feasible schedules that dominates λ, we must have

$$\sum_r \sigma_r^*(t) q_r(t)^\alpha > \sum_r \lambda_r q_r(t)^\alpha,$$

provided at least one of the queues is non-zero. This means

$$\mathbb{E}[\mathcal{L}(q(t+1)) - \mathcal{L}(q(t)) \mid q(t) \text{ large}] < 0, \quad \text{under (7.14)}. \tag{7.15}$$

In order to make the above argument rigorous, we require somewhat stronger conditions on the drift of $\mathcal{L}$. Specifically, we need to check that there exist constants $K > 0$, $\epsilon > 0$, and b for which the following holds. First, if $\|q(t)\| > K$, then $\mathcal{L}$ has downward drift that is bounded away from 0:

$$\mathbb{E}[\mathcal{L}(q(t+1)) - \mathcal{L}(q(t)) \mid \|q(t)\| > K] < -\epsilon < 0.$$

Second, if $\|q(t) \leq K\|$, then the upward jumps of $\mathcal{L}$ need to be uniformly bounded:

$$\mathbb{E}[\mathcal{L}(q(t+1)) - \mathcal{L}(q(t)) \mid \|q(t)\| \leq K] < b.$$

Foster–Lyapunov criteria (Proposition D.1) then assert that the Markov chain $q(t)$ is positive recurrent.

You will check in Exercise 7.18 that these inequalities hold for the quantity appearing on the right-hand side of (7.14). Showing that the approximation made in (7.14) doesn't destroy the inequalities requires controlling the tail of the distribution of the arrivals and services happening over a single time step; see Example D.4 and Exercise D.1. □

Remark 7.14 We have shown that when the queues $q(t)$ are large, the MaxWeight-α algorithm chooses the schedule that will maximize the one-step downward drift of $\mathcal{L}(q(t))$. Thus, it belongs to a class of greedy, or myopic, policies, which optimize some quantity over a single step. It may be surprising that we don't need to think about what happens on subsequent steps, but it certainly makes the algorithm easier to implement!

Remark 7.15 Thus arrival rates λ can be supported by a scheduling algorithm if λ is in the interior of the admissible region Λ. Since the set of schedules $\mathcal{S}$ is finite, the admissible region Λ is a polytope that can be written in the form $\Lambda = \{\lambda \geq 0 : A\lambda \leq C\}$ for some A, C with non-negative components. For the input-queued switch, Example 7.11, the linear constraints $A\lambda \leq C$ represent the $2N$ constraints identified in the representation (7.13). For this example, each constraint corresponds to the capacity of a data transmission cable connected to a port.

Note that the relevant time scales for switch scheduling are much shorter than round-trip times in a wide area network, by the ratio of the distances concerned (centimetres, rather than kilometres or thousands of kilometres). TCP's congestion control will change arrival rates at a router as a response to congestion, but this occurs over a slower time scale than that relevant for the model of this section.

Exercises

Exercise 7.17 If the arrival rate vector λ lies outside the admissible region Λ, find a vector w such that the weighted combination of workloads in the different queues, $w \cdot q$, has upwards drift whatever scheduling strategy is used.
[*Hint:* Since Λ is closed and convex, the separating hyperplane theorem shows that there is a pair of parallel hyperplanes separated by a gap in between Λ and λ.]

Exercise 7.18 Show that the right-hand side of (7.14) satisfies

$$\sum_r q_r(t)^\alpha \mathbb{E}[q_r(t+1) - q_r(t) \mid \|q(t)\| > K] < -\epsilon \|q(t)\|^\alpha$$

for some $\epsilon > 0$ and an appropriately large constant K.
[*Hint:* MaxWeight-α is invariant under rescaling all queue sizes by a constant factor, so reduce to the case $\|q(t)\| = 1$, and use the finiteness of the set $\mathcal{S}$.]

Show also that

$$\sum_r q_r(t)^\alpha \mathbb{E}[q_r(t+1) - q_r(t) \mid \|q(t)\| \le K] < b$$

for some constant b.

Example D.4 and Exercise D.1 derive similar bounds for the quantity on the left-hand side of (7.14), which are needed to apply the Foster–Lyapunov criteria.

Exercise 7.19 Suppose a wireless network is serving the seven cells illustrated in Figure 3.5, where there is a single radio channel which cannot be simultaneously used in two adjacent cells. Show that the admissible region is

$$\Lambda = \{\lambda \in [0,1]^7 : \lambda_\alpha + \lambda_\beta + \lambda_\gamma \le 1 \text{ for all } \alpha, \beta, \gamma \text{ that meet at a vertex}\}.$$

Exercise 7.20 In this section we assumed a centralized form of scheduling, so that the schedule σ can be chosen as a function of queue sizes at the stations. But this is often a quite difficult computational task: to select a maximum weight schedule is in general an NP-hard problem. In this and the following exercises we'll explore a rather different approach, based on a randomized and decentralized approximation algorithm.

Recall that we have seen a continuous time version of the wireless network of Example 7.12 in Section 5.4, operating under a simple decentralized policy. In this exercise we shall see that it is possible to tune this earlier model so that it slowly adapts its parameters to meet any arrival rate vector λ in the interior of admissible region Λ.

Suppose the state $\sigma \in \mathcal{S} \subset \{0,1\}^K$ evolves as a Markov process with transition rates as in Section 5.4, and with equilibrium distribution (5.8) determined by the vector of parameters θ. Then the proportion of time that station r is transmitting is given by expression (5.9) as

$$x_r(\theta) \equiv \frac{\sum_{\sigma \in \mathcal{S}} \sigma_r \exp(\sigma \cdot \theta)}{\sum_{m \in \mathcal{S}} \exp(m \cdot \theta)}.$$

Suppose now that for $r \in \mathcal{R}$ station r varies $\theta_r = \theta_r(t)$, but so slowly that $x_r(\theta)$ is still given by the above expression. If $\mathcal{V}(\theta)$ is defined as in Exercise 5.10, show that $d\mathcal{V}(\mu(t))/dt$ will be non-negative provided we set

$$\frac{d}{dt}\theta_r(t) \gtrless 0 \text{ according as } \lambda_r \gtrless x_r(\theta(t)).$$

Exercise 7.21 Consider the following variant of the optimization problem (5.10):

$$\begin{aligned}
&\text{maximize} && \sum_{r \in \mathcal{R}} w_r \log x_r - \sum_{n \in \mathcal{S}} p(n) \log p(n) \\
&\text{subject to} && \sum_{n \in \mathcal{S}} p(n) n_r = x_r, \quad r \in \mathcal{R}, \\
&\text{and} && \sum_{n \in \mathcal{S}} p(n) = 1 \\
&\text{over} && p(n) \geq 0, \quad n \in \mathcal{S}; \quad x_r, r \in \mathcal{R}.
\end{aligned}$$

The first term of the objective function is the usual weighted proportionally fair utility function, and the second term is, as in Section 5.4, the entropy of the probability distribution $(p(n), n \in \mathcal{S})$. Show that the Lagrangian dual can be written (after omitting constant terms) in the form

$$\begin{aligned}
&\text{maximize} \quad \mathcal{V}(\theta) = \sum_{r\in\mathcal{R}} w_r \log\theta_r - \log\left(\sum_{n\in\mathcal{S}} \exp\left(\sum_{r\in\mathcal{R}} \theta_r n_r\right)\right) \\
&\text{over} \quad \theta_r \geq 0, \quad r \in \mathcal{R}.
\end{aligned}$$

Exercise 7.22 We continue with the model of Exercise 7.20. Again suppose that for $r \in \mathcal{R}$ station r can vary θ_r, but slowly so that $x_r(\theta)$ tracks its value under the equilibrium distribution determined by θ. Specifically, suppose that

$$\frac{d}{dt}\theta_r(t) = \kappa_r \left(w_r - \theta_r(t)x_r(\theta(t))\right),$$

where $\kappa_r > 0$, for $r \in \mathcal{R}$. Show that the strictly concave function $\mathcal{V}(\theta)$ of Exercise 7.21 is a Lyapunov function for this system, and that the unique value maximizing this function is an equilibrium point of the system, to which all trajectories converge.

Exercise 7.23 Consider again Example 5.20, of wavelength routing. Suppose source–sink s varies $\theta_r = \theta_r(t)$, for $r \in s$, but so slowly that $x_r(\theta)$ tracks its value under the equilibrium distribution determined by θ, given by the expression

$$x_r(\theta) \equiv \frac{\sum_{n\in\mathcal{S}(C)} n_r \exp(n\cdot\theta)}{\sum_{m\in\mathcal{S}(C)} \exp(m\cdot\theta)}.$$

Specifically, suppose that $\theta_r(t) = \theta_s(t)$ for $r \in s$, and that

$$\frac{d}{dt}\theta_s(t) = \kappa_s \left(w_s - \theta_s(t)\sum_{r\in s} x_r(\theta(t))\right),$$

where $\kappa_s > 0$, for $s \in \mathcal{S}$.

Show that the strictly concave function of $(\theta_s, s \in \mathcal{S})$

$$\mathcal{V}(\theta) = \sum_{s\in\mathcal{S}} w_s \log\theta_s - \log\left(\sum_{n\in\mathcal{S}(C)} \exp\left(\sum_{s\in\mathcal{S}} \theta_s \sum_{r\in s} n_r\right)\right)$$

is a Lyapunov function for this system, and that the unique value maximizing this function is an equilibrium point, to which all trajectories converge.

Observe that this model gives throughputs on source–sink pairs s that are weighted proportionally fair, with weights $(w_s, s \in \mathcal{S})$.

7.9 Further reading

Srikant (2004) and Shakkottai and Srikant (2007) provide a more extensive introduction to the mathematics of Internet congestion control, and survey work up until their dates of publication. The review by Chiang *et al.* (2007) develops implications for network architecture (i.e. which functions are performed where in a network). Johari and Tsitsiklis (2004) treat the strategic behaviour of price-anticipating users, a topic touched upon in Exercise 7.1, and Berry and Johari (2013) is a recent monograph on economic models of engineered networks.

For more on the particular primal and dual models with delays of Section 7.7, see Kelly (2003a); a recent text on delay stability for networks is Tian (2012). There is much interest in the engineering and implementation of congestion control algorithms: see Vinnicombe (2002) and Kelly (2003b) for the TCP variant of Section 7.7, Kelly and Raina (2011) for a dual algorithm using the local stability condition (7.12), and Wischik *et al.* (2011) for an account of recent work on multipath algorithms.

Tassiulas and Ephremides (1992) established the stability of the maximum weight scheduling algorithm. Shah and Wischik (2012) is a recent paper on MaxWeight-α scheduling, which discusses the interesting conjecture that for the switch of Example 7.11 the average queueing delay decreases as α decreases. Our assumption, in Theorem 7.13, of Poisson arrivals is stronger than needed: Bramson (2006) shows that when a carefully defined *fluid model* of a queueing network is stable, the queueing network is positive recurrent.

Research proceeds apace on randomized and decentralized approximation algorithms for hard scheduling problems. For recent reviews, see Jiang and Walrand (2012), Shah and Shin (2012) and Chen *et al.* (2013): a key issue concerns the separation of time scales implicit in these algorithms.

8

Flow level Internet models

In the previous chapter, we considered a network with a fixed number of users sending packets. In this chapter, we look over longer time scales, where the users may leave because their files have been transferred, and new users may arrive into the system. We shall develop a stochastic model to represent the randomly varying number of flows present in a network where bandwidth is dynamically shared between flows, where each flow corresponds to the continuous transfer of an individual file or document. We assume that the rate control mechanisms we discussed in Chapter 7 work on a much faster time scale than these changes occur, so that the system reaches its equilibrium rate allocation very quickly.

8.1 Evolution of flows

We suppose that a flow is transferring a file. For example, when Elena on her home computer is downloading files from her office computer, each file corresponds to a separate flow. In this chapter, we allow the number of flows using a given route to fluctuate. Let n_r be the number of active flows along route r. Let x_r be the rate allocated to each flow along route r (we assume that it is the same for each flow on the same route); then the capacity allocated to route r is $n_r x_r$ at each resource $j \in r$. The vector $x = (x_r, r \in \mathcal{R})$ will be a function of $n = (n_r, r \in \mathcal{R})$; for example, it may be the equilibrium rate allocated by TCP, when there are n_r users on route r.

We assume that new flows arrive on route r as a Poisson process of rate ν_r, and that files transferred over route r have a size that is exponentially distributed with parameter μ_r. Assume file sizes are independent of each other and of arrivals. Thus, the number of flows evolves as follows:

$$n_r \to n_r + 1 \text{ at rate } \nu_r, \quad n_r \to n_r - 1 \text{ at rate } \mu_r n_r x_r(n),$$

for each $r \in \mathcal{R}$. With this definition, n is a Markov process.

The ratio $\rho_r := \nu_r / \mu_r$ is the *load* on route r.

Remark 8.1 This model corresponds to a *time-scale separation*: given the number of flows n, we assume the rate control mechanism instantly achieves rates $x_r(n)$. The assumption that file sizes are exponentially distributed is not an assumption likely to be satisfied in practice, but fortunately the model is not very sensitive to this assumption.

Exercises

Exercise 8.1 Suppose the network comprises a single resource, of capacity C, and that $x_r(n) = C/N$, where $N = \sum_r n_r$, so that the capacity is shared equally over all file transfers in progress. Show that, provided

$$\rho := \sum_{r \in \mathcal{R}} \rho_r < C,$$

the Markov process $n = (n_r, r \in \mathcal{R})$ has equilibrium distribution

$$\pi(n) = \left(1 - \frac{\rho}{C}\right) \binom{N}{n_1, n_2, \ldots, n_{|\mathcal{R}|}} \prod_{r \in \mathcal{R}} \left(\frac{\rho_r}{C}\right)^{n_r},$$

and deduce that in equilibrium N has a geometric distribution with mean C/ρ. Observe that if $\rho \geq C$, then omitting the term $(1 - \rho/C)$ in the above expression for π still gives a solution to the equilibrium equations, but now its sum is infinite and so no equilibrium distribution exists (if $\rho = C$ the process is null recurrent, and if $\rho > C$ the process is transient).

Show that the mean file size over all files arriving to be transferred is $\rho / \sum_r \nu_r$. Deduce that when file sizes arriving at the resource are distributed as a mixture of exponential random variables, the distribution of N depends only on the mean file size, and not otherwise on the distribution of file sizes.

Exercise 8.2 Suppose the shared part of the network comprises a single resource of capacity C, but that each flow has to pass through its own access link of capacity L as well, so that $x_r(n) = \max\{L, C/N\}$, where N is the number of flows present. Suppose flows arrive as a Poisson process and files are exponentially distributed. Show that if $C = sL$ then the number of flows in the system evolves as an $M/M/s$ queue.

8.2 α-fair rate allocations

Next we define the rate allocation $x = x(n)$. We define a family of allocation policies that includes as special cases several examples that we have seen earlier.

Let $w_r > 0, r \in \mathcal{R}$, be a set of weights, and let $\alpha \in (0, \infty)$ be a fixed constant. The *weighted α-fair allocation* $x = (x_r, r \in \mathcal{R})$ is, for $\alpha \neq 1$, the solution of the following optimization problem:

$$\begin{aligned} &\text{maximize} \sum_r w_r n_r \frac{x_r^{1-\alpha}}{1-\alpha} \\ &\text{subject to} \sum_{r:j\in r} n_r x_r \leq C_j, \qquad j \in \mathcal{J}, \\ &\text{over} \qquad x_r \geq 0, \qquad r \in \mathcal{R}. \end{aligned} \tag{8.1}$$

For $\alpha = 1$, it is the solution of the same optimization problem but with objective function $\sum_r w_r n_r \log x_r$; Exercise 8.3 shows why this is the natural continuation of the objective function through $\alpha = 1$.

For all $\alpha \in (0, \infty)$ and all n, the objective function is a strictly concave function of $(x_r : r \in \mathcal{R}, n_r > 0)$; the problem is a generalization of NETWORK$(A, C; w)$, from Section 7.1.

We can characterize the solutions of the problem (8.1) as follows. Let $(p_j, j \in \mathcal{J})$ be Lagrange multipliers for the constraints. Then, at the optimum, $x = x(n)$ and $p = p(n)$ satisfy

$$x_r = \left(\frac{w_r}{\sum_{j\in r} p_j} \right)^{1/\alpha}, \qquad r \in \mathcal{R},$$

where

$$x_r \geq 0, r \in \mathcal{R}, \quad \sum_{r:j\in r} n_r x_r \leq C_j, \qquad j \in \mathcal{J} \quad \text{(primal feasibility);}$$

$$p_j \geq 0, \qquad j \in \mathcal{J} \quad \text{(dual feasibility);}$$

$$p_j \cdot \left(C_j - \sum_{r:j\in r} n_r x_r \right) = 0, \qquad j \in \mathcal{J} \quad \text{(complementary slackness).}$$

The form of an α-fair rate allocation captures several of the fairness definitions we have seen earlier.

- As $\alpha \to 0$ and with $w_r \equiv 1$ the total throughput, $\sum_r n_r x_r$, approaches its maximum.
- If $\alpha = 1$ then, from Proposition 7.4, the rates x_r are weighted proportionally fair.
- If $\alpha = 2$ and $w_r = 1/T_r^2$ then the rates x_r are *TCP fair*, i.e. they are of the form (7.9).
- As $\alpha \to \infty$ and with $w_r \equiv 1$ the rates x_r approach max-min fairness.

Exercises

Exercise 8.3 Check that the objective function of the problem (8.1) is concave for each of the three cases $0 < \alpha < 1, \alpha = 1$ and $1 < \alpha < \infty$. Show that the derivative of the objective function with respect to x_r has the same form, namely $w_r n_r x_r^{-\alpha}$, for all $\alpha \in (0, \infty)$.

Exercise 8.4 Check the claimed characterization of the solutions to the problem (8.1).

Exercise 8.5 Consider the α-fair allocation as $\alpha \to 0$, the allocation maximizing the total throughput. Show that however large the number of distinct routes $|\mathcal{R}|$, typically only $|\mathcal{J}|$ of them are allocated a non-zero rate. [*Hint:* Show that as $\alpha \to 0$ the problem (8.1) approaches a linear program, and consider basic feasible solutions.]

8.3 Stability of α-fair rate allocations

Consider now the behaviour of the Markov process n under an α-fair allocation. Is it stable? This will clearly depend on the parameters ν_r, μ_r, and we show next that the condition is as simple as it could be: all that is needed is that the load arriving at the network for resource j is less than the capacity of resource j, for every resource j.

Theorem 8.2 *The Markov process n is positive recurrent (i.e. has an equilibrium distribution) if and only if*

$$\sum_{r:j\in r} \rho_r < C_j \quad \textit{for all } j \in J. \tag{8.2}$$

Sketch of proof If condition (8.2) is violated for some j then the system is not stable, by the following coupling argument. Suppose that j is the only resource with finite capacity in the network, with all the other resources given infinite capacity. This cannot increase the work that has been processed at resource j before time t, for any $t \geq 0$. But the resulting system would be the single server queue considered in Exercise 8.1, which is transient or at best null recurrent.

Our approach to showing that under condition (8.2) n is positive recurrent will be the same as in the proof of Theorem 7.13. We shall write down a Lyapunov function, approximate its drift, and show that the approximation is negative when n is large, and bounded when n is small. This is almost enough to apply the Foster–Lyapunov criteria (Proposition D.3); you will check the remaining details in Exercise D.2.

We begin by calculating the drift of n.

$$\mathbb{E}[n_r(t+\delta t) - n_r(t) \,|\, n(t)] \approx \big(\nu_r - \mu_r n_r x_r(n(t))\big)\delta t.$$

This would be an equality if we could guarantee that not all of the $n_r(t)$ flows depart by time $t + \delta t$. As $n_r(t) \to \infty$, the approximation improves.

Next, we consider the Lyapunov function

$$\mathcal{L}(n) = \sum_r \frac{w_r}{\mu_r} \rho_r^{-\alpha} \frac{n_r^{\alpha+1}}{\alpha+1}.$$

This is a non-negative function. We argue that the drift of $\mathcal{L}$ should be negative provided $n(t)$ is large. We can approximate the change of $\mathcal{L}$ over a small time period by

$$\begin{aligned}\frac{1}{\delta t}\mathbb{E}[\mathcal{L}(n(t+\delta t)) - \mathcal{L}(n(t)) \,|\, n(t)] &\approx \sum_{r\in\mathcal{R}} \left(\frac{\partial \mathcal{L}}{\partial n_r}\right) \cdot \frac{1}{\delta t}\mathbb{E}[n_r(t+\delta t) - n_r(t) \,|\, n(t)] \\ &= \sum_r \frac{w_r}{\mu_r} \rho_r^{-\alpha} n_r^{\alpha}\big(\nu_r - \mu_r n_r x_r(n(t))\big) \\ &= \sum_r w_r \rho_r^{-\alpha} n_r^{\alpha}\big(\rho_r - n_r x_r(n(t))\big). \end{aligned} \tag{8.3}$$

Our goal is to show that this is negative.

At this point it is helpful to rewrite the optimization problem (8.1) in terms of $X_r = n_r x_r$, $r \in \mathcal{R}$, as follows:

$$\begin{aligned} &\text{maximize } G(X) = \sum_r w_r n_r^{\alpha} \frac{X_r^{1-\alpha}}{1-\alpha} \\ &\text{subject to } \sum_{r:j\in r} X_r \le C_j, \qquad j \in \mathcal{J}, \\ &\text{over} \qquad X_r \ge 0, \qquad r \in \mathcal{R}. \end{aligned} \tag{8.4}$$

Since G is concave, for every U inside the feasible region of (8.4)

$$G'(U) \cdot (U - X) \le G(U) - G(X) \le 0, \tag{8.5}$$

where X is the optimum (Figure 8.1 illustrates this).

Now, if the stability conditions (8.2) are satisfied, then $\exists \epsilon > 0$ such that $u = (\rho_r(1+\epsilon),\ r \in \mathcal{R})$ is also inside the region (8.2). Therefore, by (8.5),

$$\sum_r w_r n_r^{\alpha} (\rho_r(1+\epsilon))^{-\alpha} (\rho_r(1+\epsilon) - X_r) \le 0.$$

Combining this with (8.3), we see that, for large n, the drift of $\mathcal{L}(n)$ is

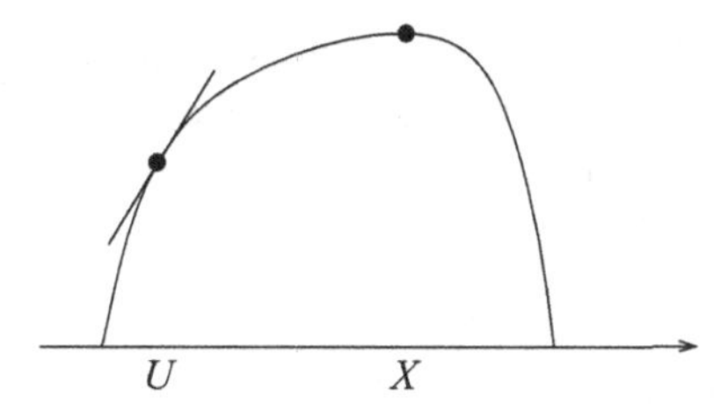

Figure 8.1 Let X be the optimum and let U be another point in the feasible region of (8.4). Since $G(\cdot)$ is concave, the tangent plane at U lies above $G(\cdot)$, so $G(X) \leq G(U) + G'(U) \cdot (X - U)$.

negative:

$$\frac{1}{\delta t}\mathbb{E}[\mathcal{L}(n(t+\delta t)) - \mathcal{L}(n(t)) \,|\, n(t)] \leq -\epsilon \sum_r w_r n_r^{\alpha} \rho_r^{-\alpha+1}.$$

In order to apply the Foster–Lyapunov criteria (Proposition D.1), we also need to bound the drift of $\mathcal{L}(n)$ from above. You will do this (for the quantity on the right-hand side of (8.3)) in Exercise 8.6. This can be done using the fact that flows arrive as Poisson processes, and depart at most as quickly as a Poisson process. □

Remark 8.3 It is interesting to compare the model of an α-fair rate allocation with the earlier model of a MaxWeight-α queue schedule. The models are operating on very different time scales (the time taken for a packet to pass through a router, versus the time taken for a file to pass across a network), but nevertheless the stability results established are of the same form.

If we view n_r as the size of the queue of files that are being transmitted along route r, then an α-fair rate allocation chooses the rates X_r allocated to these queues to solve the optimization problem (8.4). A MaxWeight-α schedule would instead choose the rates X_r allocated to these queues to solve the problem

$$\begin{aligned}
&\text{maximize} && \sum_r n_r^{\alpha} X_r \\
&\text{subject to} && \sum_{r:j\in r} X_r \leq C_j, \qquad j \in \mathcal{J}, \\
&\text{over} && X_r \geq 0, \qquad r \in \mathcal{R}.
\end{aligned}$$

In either case we observe that if a queue n_r grows to infinity while other queues are bounded, then this must eventually cause the rate X_r allocated to this queue to increase towards its largest feasible value. In Section 8.4 we consider a rate allocation scheme for which this is not the case.

Exercise

Exercise 8.6 Show that the right-hand side of (8.3) satisfies

$$\sum_r \frac{w_r}{\mu_r} \rho_r^{-\alpha} n_r^{\alpha} \frac{1}{\delta t} \mathbb{E}[n_r(t+\delta t) - n_r(t) \mid \|n(t)\| \le K] \le b$$

for some constant b. Why does that not follow immediately from the calculation in the proof of Theorem 8.2?

Exercise D.2 in the Appendix asks you to prove a similar inequality for the quantity on the left-hand side of (8.3), which is needed to apply the Foster–Lyapunov criteria (Proposition D.3).

8.4 What can go wrong?

The stability condition (8.2) is so natural that it might be thought that it would apply to most rate allocation schemes. In this section we show otherwise.

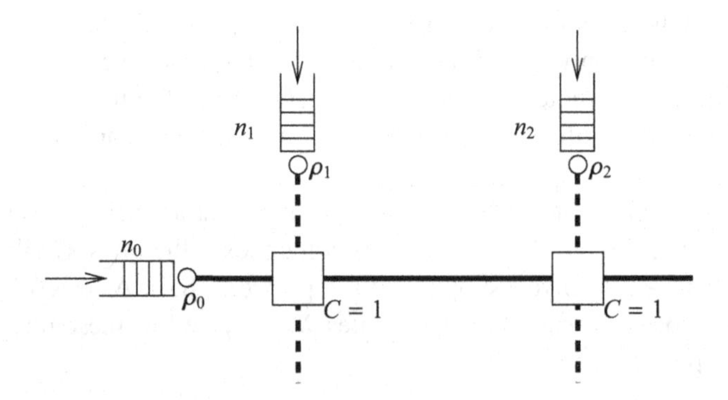

Figure 8.2 A system with three flows and two resources.

Consider the network illustrated in Figure 8.2 with just two resources and three routes: two single-resource routes and one route using both resources. With any α-fair allocation of the traffic rates, as we saw in Section 8.3, the stability condition is $\rho_0 + \rho_1 < 1$, $\rho_0 + \rho_2 < 1$. Provided this condition is satisfied, the Markov process describing the number of files on each route is positive recurrent.

Let us consider a different way of allocating the available capacity. Suppose that streams 1 and 2 are given absolute priority at their resources: that is, if $n_1 > 0$ then $n_1 x_1 = 1$ (and hence $n_0 x_0 = 0$), and if $n_2 > 0$ then $n_2 x_2 = 1$ (and again $n_0 x_0 = 0$). Then stream 0 will only get served if $n_1 = n_2 = 0$, i.e. if there is no work in any of the high-priority streams.

What is the new stability condition? Resource 1 is occupied by stream 1 a proportion ρ_1 of the time; independently, resource 2 is occupied by stream 2 a proportion ρ_2 of the time. (Because of the absolute priority, neither stream 1 nor stream 2 "sees" stream 0.) Therefore, both of these resources are free for stream 0 to use a proportion $(1 - \rho_1)(1 - \rho_2)$ of the time, and the stability condition is thus

$$\rho_0 < (1 - \rho_1)(1 - \rho_2).$$

This is a strictly smaller stability region than before. Figure 8.3 shows a slice of the stability region for a fixed value of ρ_0.

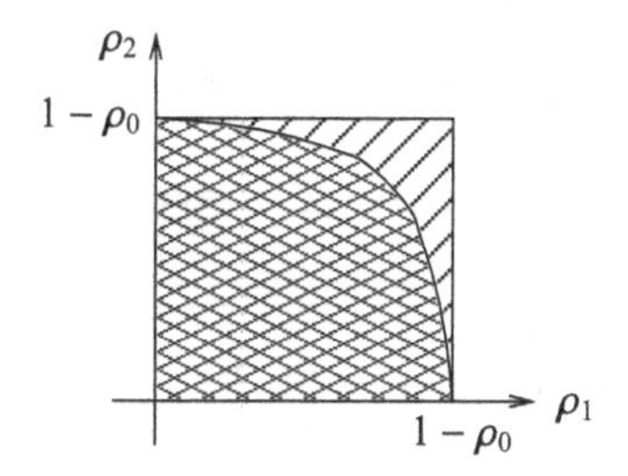

Figure 8.3 Stability regions under α-fair and priority schemes.

The system is failing to realize its capacity because there is *starvation* of resources: the high-priority flows prevent work from reaching some of the available resources.

Remark 8.4 Suppose that we put a very large buffer between resources 1 and 2, allowing jobs on route 0 to be served by resource 1 and then wait for service at resource 2. This would increase the stability region, and might be an adequate solution in some contexts. However, in the Internet the buffer sizes required at internal nodes would be unrealistically large, and additional buffering between resources would increase round-trip times.

Another way of giving priority to customers on routes 1 and 2 would be to use the weighted α-fair scheme and assign these flows very high weight;

as we know, any weighted α-fair scheme results in a positive recurrent Markov process within the larger stability region. Why are the results for the two schemes so different? In both schemes, the number of files on route 0, n_0, will eventually become very large. In an α-fair scheme, that means that route 0 will eventually receive service; in the strict priority setting, n_0 doesn't matter at all.

Strict priority rules often cause starvation of network resources and a reduction in the stability region. Exercise 8.7 gives a queueing network example.

Exercise

Exercise 8.7 Consider the network in Figure 8.4. Jobs in the network need to go through stations 1–4 in order, but a single server is working at stations 1 and 4, and another single server is working at stations 2 and 3. Consequently, at any time, either jobs at station 1, or jobs at station 4, but not both, may be served; similarly for jobs at stations 2 and 3.

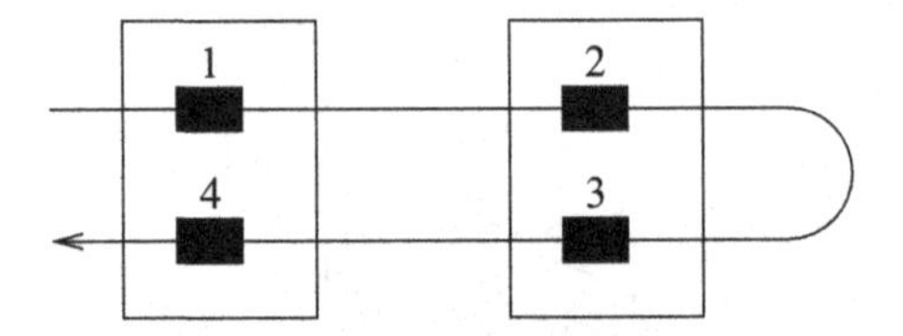

Figure 8.4 The Lu–Kumar network. Jobs go through stations 1–4 in order, but a single server is working at stations 1 and 4, and at 2 and 3.

We will assume that jobs at station 4 have strict priority over station 1, and that jobs at station 2 have strict priority over station 3. That is, if the server at stations 1 and 4 can choose a job at station 1 or at station 4, she will process the job at station 4 first, letting the jobs at station 1 queue.

Assume that jobs enter the system deterministically, at times 0, 1, 2, and so on. We also assume that the service times of all the jobs are deterministic: a job requires time 0 at the low-priority stations 1 and 3, and time 2/3 at the high-priority stations 2 and 4. The zero-time jobs are, of course, an idealization, but we'll see that they still make a difference due to the priority rules.

Assume that service completions happen "just before schedule". In particular, a job completing service at station 3 at time n will arrive at station 4 before the external arrival at time n comes to station 1.

(1) Check that each of the two servers individually is not overloaded, i.e. work is arriving at the system for each of them at a rate smaller than 1.
(2) Let M be a large integer divisible by 3, and consider this system starting at time 0 with M jobs waiting to be processed by station 1, and no jobs anywhere else. At time 0, these jobs all move to station 2. What is the state of the system at time $2M/3$? At time M? When will the system first start processing jobs at station 3?
(3) Once the jobs are processed through station 3, they move on to station 4, which has priority over station 1. When will jobs be processed next through station 1? What is the queue at station 1 at that time?

This example shows that, if we start the system at time 0 with a queue of M jobs waiting for service at station 1, then at time $4M$ there will be a queue of $2M$ jobs there. That is, the system is not stable. This is happening because the priority rules cause starvation of some of the servers, forcing them to be idle for a large portion of the time.

8.5 Linear network with proportional fairness

In this section, we analyze an example of a simple network with a proportionally fair rate allocation. A linear network (illustrated in Figure 8.5) is in some respects what a more complicated network looks like from the point of view of a single file; consequently, it is an important model to understand. We shall see that it is possible to find its equilibrium distribution explicitly, at least for certain parameter values.

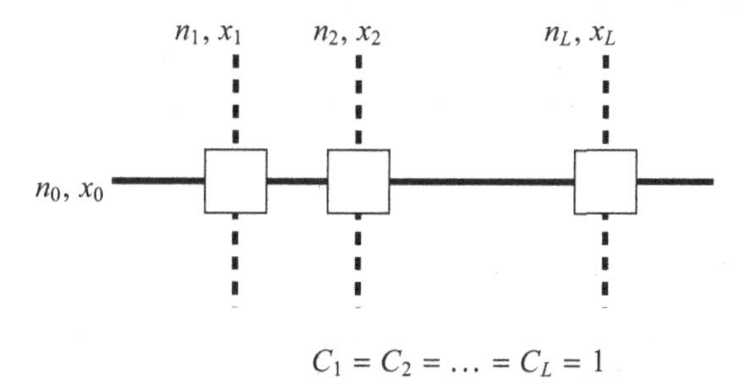

Figure 8.5 Linear network with L resources and $L + 1$ flows.

Take $\alpha = 1$ and $w_i = 1$ for all i, i.e. proportional fairness. Let us compute the proportionally fair rate allocation by maximizing $\sum n_i \log x_i$ over the feasible region for x. Clearly, if $n_l > 0$ then $n_0 x_0 + n_l x_l = 1$, as otherwise we can increase x_l. Therefore, the optimization problem requires that we maximize

$$n_0 \log x_0 + \sum_{l=1}^{L} n_l \log\left(\frac{1 - n_0 x_0}{n_l}\right),$$

where the summation runs over l for which $n_l > 0$. Differentiating, at the optimum we have

$$\frac{n_0}{x_0} = \sum_{l=1}^{L} \frac{n_l n_0}{1 - n_0 x_0} \implies 1 - n_0 x_0 = x_0 \sum_{l=1}^{L} n_l$$

or

$$x_0 = \frac{1}{n_0 + \sum_{l=1}^{L} n_l}; \qquad n_0 x_0 = \frac{n_0}{n_0 + \sum_{l=1}^{L} n_l} = 1 - n_l x_l.$$

We now compute the equilibrium distribution for this system explicitly. The general formulae for transition rates are

$$q(n, n + e_r) = \nu_r, \qquad q(n, n - e_r) = x_r(n) n_r \mu_r,$$

and thus in this example

$$q(n, n - e_0) = \mu_0 \frac{n_0}{n_0 + \sum_{l=1}^{L} n_l}, \qquad n_0 > 0,$$

and

$$q(n, n - e_i) = \mu_i \frac{\sum_{l=1}^{L} n_l}{n_0 + \sum_{l=1}^{L} n_l}, \qquad n_i > 0,\ i = 1, \ldots, L.$$

Theorem 8.5 *The stationary distribution for the above network is*

$$\pi(n) = \frac{\prod_{l=1}^{L}(1 - \rho_0 - \rho_l)}{(1 - \rho_0)^{L-1}} \binom{\sum_{l=0}^{L} n_l}{n_0} \prod_{l=0}^{L} \rho_l^{n_l}$$

provided $\rho_0 + \rho_l < 1$ *for* $l = 1, 2, \ldots, L$.

Proof We check the detailed balance equations,

$$\pi(n) \underbrace{q(n, n + e_0)}_{\nu_0} = \underbrace{\pi(n + e_0)}_{\pi(n) \frac{n_0 + \sum_{l=1}^{L} n_l + 1}{n_0 + 1} \frac{\nu_0}{\mu_0}} \underbrace{q(n + e_0, n)}_{\mu_0 \frac{n_0 + 1}{n_0 + \sum_{l=1}^{L} n_l + 1}}$$

and

$$\pi(n) \underbrace{q(n, n + e_i)}_{\nu_i} = \underbrace{\pi(n + e_i)}_{\pi(n) \frac{n_0 + \sum_{l=1}^{L} n_l + 1}{\sum_{l=1}^{L} n_l + 1} \frac{\nu_i}{\mu_i}} \underbrace{q(n + e_i, n)}_{\mu_i \frac{\sum_{l=1}^{L} n_l + 1}{n_0 + \sum_{l=1}^{L} n_l + 1}},$$

as required.

It remains only to check that π sums to 1. But

$$\sum_{n_0,n_1,\ldots,n_L} \binom{n_0+n_1+\ldots+n_L}{n_0} \rho_0^{n_0}\rho_1^{n_1}\cdots\rho_L^{n_L}$$

$$= \sum_{n_1,\ldots,n_L} \rho_1^{n_1}\cdots\rho_L^{n_L} \sum_{n_0} \binom{n_0+\ldots+n_L}{n_0} \rho_0^{n_0}$$

$$= \sum_{n_1,\ldots,n_L} \rho_1^{n_1}\cdots\rho_L^{n_L} \frac{1}{(1-\rho_0)^{n_1+\ldots+n_L+1}} \qquad \text{(negative binomial expansion)}$$

$$= \frac{1}{1-\rho_0} \sum_{n_1,\ldots,n_L} \left(\frac{\rho_1}{1-\rho_0}\right)^{n_1} \left(\frac{\rho_2}{1-\rho_0}\right)^{n_2} \cdots \left(\frac{\rho_L}{1-\rho_0}\right)^{n_L} \qquad \text{if } \rho_i < 1-\rho_0,\ \forall i$$

$$= \frac{(1-\rho_0)^{L-1}}{\prod_{l=1}^{L}(1-\rho_0-\rho_l)},$$

and thus π does indeed sum to 1, provided the stability conditions are satisfied. □

Remark 8.6 The above network is an example of a quasi-reversible queue. That is, if we draw a large box around the linear network, then arrivals of each file type will be Poisson processes by assumption, and the departures will be Poisson as well. (We saw this for a sequence of $M/M/1$ queues in Chapter 2.) This lets us embed the linear network into a larger network as a unit, in the same way we did with $M/M/1$ queues. We could, for example, replace the simple processor sharing queue 0 in Figure 2.9 by the above linear network and retain a product form. It also can be used to show insensitivity, i.e. that the stationary distribution depends only on the mean file sizes, and not otherwise on their distributions.

Remark 8.7 We can compute the average time to transfer a file on route r but, as files on different routes may have very different sizes, a more useful measure is the *flow throughput* on route r, defined as

$$\frac{\text{average file size on route } r}{\text{average time required to transfer a file on route } r}.$$

We know that the average file size is $1/\mu_r$. The average transfer time can be derived from Little's law, $L = \lambda W$, which tells us

$$\mathbb{E}(n_r) = \nu_r \cdot \text{average time required to transfer a file on route } r.$$

Combining these, and recalling $\rho_r = \nu_r/\mu_r$, we obtain

$$\text{flow throughput} = \frac{\rho_r}{\mathbb{E}n_r}.$$

Since we know $\pi(n)$, we can calculate this throughput, at least for the particular model of this section: see Exercise 8.8.

Exercises

Exercise 8.8 Show that the flow throughput, $\rho_r/\mathbb{E}n_r$, for the linear network is given by

$$\begin{cases} 1 - \rho_0 - \rho_l, & l = 1, \dots, L, \\ (1-\rho_0)\left(1 + \sum_{l=1}^{L} \frac{\rho_0}{1 - \rho_0 - \rho_l}\right)^{-1}, & l = 0. \end{cases}$$

Note that $1 - \rho_0 - \rho_l$ is the proportion of time that resource l is idle. Thus, the flow throughput on route $l = 1, \dots, L$ is as if a file on this route gets a proportion $1 - \rho_0 - \rho_l$ of resource l to itself.

Exercise 8.9 Use the stationary distribution found in Theorem 8.5 to show that under this distribution $n_1, n_2, \dots, n_L$ are independent, and n_l is geometrically distributed with mean $\rho_l/(1 - \rho_0 - \rho_l)$.
[*Hint:* Sum $\pi(n)$ over n_0, and use the negative binomial expansion.]

Exercise 8.10 Consider a network with resources $\mathcal{J} = \{1, 2, 3, 4\}$, each of unit capacity, and routes $\mathcal{R} = \{12, 23, 34, 41\}$, where we use ij as a convenient shorthand for $\{i, j\}$. Given $n = (n_r, r \in \mathcal{R})$, find the rate x_r of each flow on route r, for each $r \in \mathcal{R}$, under a proportionally fair rate allocation. Show, in particular, that if $n_{12} > 0$ then

$$x_{12}n_{12} = \frac{n_{12} + n_{34}}{n_{12} + n_{23} + n_{34} + n_{41}}.$$

Suppose now that flows describe the transfer of files through a network, that new flows originate as independent Poisson processes of rates $\nu_r, r \in \mathcal{R}$, and that file sizes are independent and exponentially distributed with mean μ_r on route $r \in \mathcal{R}$. Determine the transition rates of the resulting Markov process $n = (n_r, r \in \mathcal{R})$. Show that the stationary distribution of the Markov process $n = (n_r, r \in \mathcal{R})$ takes the form

$$\pi(n) = B\binom{n_{12} + n_{23} + n_{34} + n_{41}}{n_{12} + n_{34}} \prod_r \left(\frac{\nu_r}{\mu_r}\right)^{n_r},$$

provided it exists.

8.6 Further reading

Theorem 8.2 and the examples of Sections 8.4 and 8.5 are from Bonald and Massoulié (2001); for further background to the general approach of this chapter, see BenFredj *et al.* (2001).

The network in Exercise 8.7 was introduced by Lu and Kumar (1991). Analyzing the set of arrival rates for which a given queueing network with a given scheduling discipline will be stable turns out to be a surprisingly non-trivial problem. The book by Bramson (2006) discusses in detail a method of showing stability using fluid models, which generalize the drift analysis we have seen in several chapters.

Kang *et al.* (2009) show that a generalization of the independence result of Exercise 8.9 holds approximately for a wide class of networks operating under proportional fairness.

In this chapter we have assumed a time-scale separation of packet-level dynamics, such as those described in Section 7.8 for MaxWeight scheduling, and flow-level dynamics, such as those described in Section 8.1. See Walton (2009) and Moallemi and Shah (2010) for more on the joint dynamics of these two time scales. Shah *et al.* (2012) is a recent paper exploring connections between scheduling strategies, such as those in Section 7.8, and rate allocations, such as those in Section 8.2.

Appendix A

Continuous time Markov processes

Let $\mathcal{T}$ be a subset of $\mathbb{R}$. A collection of random variables $(X(t), t \in \mathcal{T})$ defined on a common probability space and taking values in a countable set $\mathcal{S}$ has the Markov property if, for $t_0 < t_1 < \ldots < t_n < t_{n+1}$ all in $\mathcal{T}$,

$$\begin{aligned}\mathbb{P}(X(t_{n+1}) = x_{n+1} \,|\, X(t_n) = x_n, X(t_{n-1}) = x_{n-1}, \ldots, X(t_0) = x_0)\\ = \mathbb{P}(X(t_{n+1} = x_{n+1} \,|\, X(t_n) = x_n)\end{aligned}$$

whenever the event $\{X(t_n) = x_n\}$ has positive probability. This is equivalent to the property that, for $t_0 < t_1 < \ldots < t_n$ all in $\mathcal{T}$ and for $0 < p < n$,

$$\begin{aligned}&\mathbb{P}(X(t_i) = x_i, 0 \leq i \leq n \,|\, X(t_p) = x_p)\\ &\quad = \mathbb{P}(X(t_i) = x_i, 0 \leq i \leq p \,|\, X(t_p) = x_p)\mathbb{P}(X(t_i) = x_i, p \leq i \leq n \,|\, X(t_p) = x_p)\end{aligned}$$

whenever the event $\{X(t_p) = x_p\}$ has positive probability. Note that this statement has no preferred direction of time: it states that, conditional on $X(t_p)$ (the present), $(X(t_i), 0 \leq i \leq p)$ (the past) and $(X(t_i), p \leq i \leq n)$ (the future) are independent.

Our definition in Chapter 1 of a Markov chain corresponds to the case where $\mathcal{T}$ is $\mathbb{Z}_+$ or $\mathbb{Z}$, and where the transition probabilities are time homogeneous. Our definition of a Markov process corresponds to the case where $\mathcal{T}$ is $\mathbb{R}_+$ or $\mathbb{R}$, where the process is time homogeneous, and where there is an additional assumption which we now explore.

A possibility that arises in continuous, but not in discrete, time is *explosion*. Consider a Markov process that spends an exponentially distributed time, with mean $(1/2)^k$, in state k before moving to state $k+1$, and suppose the process starts at time 0 in state 0. Since $\sum_k (1/2)^k = 2$, we expect the process to have run out instructions by time 2. The trajectories will look something like Figure A.1. In particular, there will be an infinite number of transitions during a finite time period.

A related possibility is a process "returning from ∞": consider a process that just before time 0 is in state 0, then "jumps to ∞", and then runs a

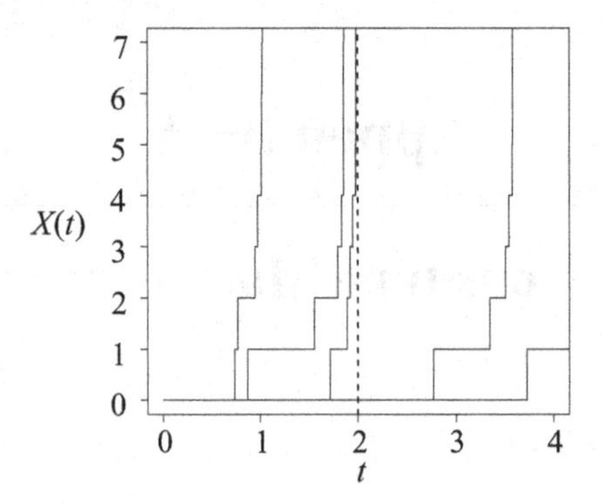

Figure A.1 Trajectories of an explosive process. On average, the time when the process "runs out of instructions" is 2.

time-reversed version of the above explosion. Then, at any time $t > 0$, the process is (with probability 1) in a finite state; but, as $t \to 0$ from above, $\mathbb{P}(X(t) = j) \to 0$ for any finite j. In particular, this process spends a finite time in state 0, but "goes nowhere" once it leaves state 0 (in this example, $q(0, j) = 0$ for all $j > 0$). Such a process is known as *non-conservative*.

Throughout this book we assume that any Markov process with which we deal remains in each state for a positive period of time and is incapable of passing through an infinite number of states in a finite time. This assumption excludes both of the above phenomena and ensures the process can be constructed as in Chapter 1 (from a jump chain and a sequence of independent exponential random variables) with $q(i) \equiv \sum_{j \in \mathcal{S}} q(i, j)$ the transition rate out of state i. It will usually be clear that the assumption holds for particular processes, for example from the definition of the process in terms of arrival and service times.

In Chapter 1, we defined an equilibrium distribution for a chain or process to be a collection $\pi = (\pi(j)), j \in \mathcal{S})$ of positive numbers summing to unity that satisfy the equilibrium equations (1.1) or (1.2). An equilibrium distribution exists if and only if the chain or process is *positive recurrent*, i.e. from any state the mean time to return to that state after leaving it is finite. If an equilibrium distribution does not exist, the chain or process may be either *null recurrent*, i.e. the time to return to a state is finite but has infinite mean, or *transient*, i.e. on leaving a state there is a positive probability that the state will never be visited again.

When an equilibrium distribution exists, the process can be explicitly constructed for $t < 0$ by constructing the reversed process $(X(-t), t \geq 0)$, using the transition rates $(q'(j,k), j,k \in \mathcal{S})$ given in Proposition 1.1. For example, in Figure 2.1, given the number in the queue at time $t = 0$, we can use the Poisson point processes A and D to construct the process both before and after time 0. This provides an alternative to starting the process with the equilibrium distribution at time t_0, and letting $t_0 \to -\infty$.

Consider a Markov process $(X(t), t \in \mathcal{T})$, where $\mathcal{T}$ is $\mathbb{R}_+$ or $\mathbb{R}$. A random variable T is a *stopping time* (for the process) if for each $t \in \mathcal{T}$ the event $\{T \leq t\}$ depends only on $(X(s), s \leq t)$; that is, it is possible to determine whether $T \leq t$ based only on observing the sample path of X up through time t. A fundamental result is the *strong Markov property*: if T is a stopping time then conditional on $T < \infty$ and $X(T) = x_p$, the past $(X(t), t \leq T)$ and the future $(X(t), t \geq T)$ are independent. If T is deterministic, this becomes our earlier restatement of the Markov property: for fixed t_0 the distribution of $(X(t), t \geq t_0)$ is determined by and determines the finite dimensional distributions, i.e. the distributions of $(X(t_i), 0 \leq i \leq n)$ for all $t_0 < t_1 < \ldots < t_n$, and any finite n (Norris (1998), Section 6). Finally, we note that our construction of a Markov process from a jump chain followed the usual convention and made the process right continuous, but this was arbitrary, and is a distinction not detected by the finite-dimensional distributions.

Appendix B

Little's law

Let $X_1, X_2, \ldots$ be independent identically distributed non-negative random variables. These will represent times between successive renewals. Let

$$N(t) = \max\{n : X_1 + \ldots + X_n \leq t\}$$

be the number of renewals that have occurred by time t; $N(t)$ is called a *renewal process*.

Suppose that a reward Y_n is earned at the time of the nth renewal. The reward Y_n will usually depend on X_n, but we suppose the pairs $(X_n, Y_n), n = 1, 2, \ldots$, are independent and identically distributed, and that Y_n, as well as X_n, is non-negative. The *renewal reward process*

$$Y(t) = \sum_{n=1}^{N(t)} Y_n$$

is the total reward earned by time t.

Theorem B.1 (Renewal reward theorem) *If both $\mathbb{E}X_n$ and $\mathbb{E}Y_n$ are finite, then*

$$\lim_{t\to\infty} \frac{Y(t)}{t} \overset{w.p.1}{=} \frac{\mathbb{E}Y_n}{\mathbb{E}X_n} = \lim_{t\to\infty} \mathbb{E}\frac{Y(t)}{t}.$$

Remark B.2 In some applications, the reward may be earned over the course of the cycle, rather than at the end. Provided the partial reward earned over any interval is non-negative, the theorem still holds. The theorem also holds if the initial pair (X_1, Y_1) has a different distribution from later pairs, provided its components have finite mean. For example, if the system under study is a Markov process with an equilibrium distribution, the system might start in equilibrium.

The reader will note the similarity of the "with probability 1" part of the theorem to the ergodic property of an equilibrium distribution for a Markov process: both results are consequences of the strong law of large numbers.

Let's prove just this part of the renewal reward theorem (the "expectation" part needs a little more renewal theory).

Proof (of the w. p. 1 part) Write (X, Y) for a typical pair (X_n, Y_n); we allow (X_1, Y_1) to have a different distribution. With probability 1, $N(t) \to \infty$ as $t \to \infty$, since the Xs have finite mean and are therefore proper random variables. Now

$$\sum_{n=1}^{N(t)} X_n \le t < \sum_{n=1}^{N(t)+1} X_n$$

by the definition of $N(t)$, and so

$$\underbrace{\frac{\sum_{n=1}^{N(t)} X_n}{N(t)}}_{\to \mathbb{E}X} \le \frac{t}{N(t)} < \underbrace{\frac{\sum_{n=1}^{N(t)+1} X_n}{N(t)+1}}_{\to \mathbb{E}X} \cdot \underbrace{\frac{N(t)+1}{N(t)}}_{\to 1}$$

where the convergences indicated are with probability 1. Thus $t/N(t) \to \mathbb{E}X$ with probability 1, and so $N(t)/t \to 1/\mathbb{E}X$ with probability 1. But

$$\sum_{n=1}^{N(t)} Y_n \le Y(t) \le \sum_{n=1}^{N(t)+1} Y_n,$$

and so

$$\underbrace{\frac{\sum_{n=1}^{N(t)} Y_n}{N(t)}}_{\to \mathbb{E}Y} \cdot \underbrace{\frac{N(t)}{t}}_{\to 1/\mathbb{E}X} \le \frac{Y(t)}{t} \le \underbrace{\frac{\sum_{n=1}^{N(t)+1} Y_n}{N(t)+1}}_{\to \mathbb{E}Y} \cdot \underbrace{\frac{N(t)+1}{N(t)}}_{\to 1} \cdot \underbrace{\frac{N(t)}{t}}_{\to 1/\mathbb{E}X},$$

and thus $Y(t)/t \to \mathbb{E}Y/\mathbb{E}X$ with probability 1. □

To deduce Little's law in Section 2.5, we needed to make three applications of the renewal reward theorem. Consider first a renewal reward process in which each customer generates unit reward per unit of time it spends in the system. Then we used this process, together with the above theorem, to define L in terms of the mean rewards earned over, and the mean length of, a regenerative cycle. Our definition of the mean arrival rate λ used a different reward structure, in which each customer generates unit reward when it enters the system. Finally, to define W we construct a discrete time renewal process, as follows. Observe the number of customers in the system at the time of each customer arrival and each customer departure. The first customer to enter the system following a regeneration point of the system marks a regeneration point on this discrete time scale. If each customer generates a reward on leaving the system equal to its time spent in the system, we define W in terms of the mean reward earned over, and the mean length of, a regenerative cycle on the discrete time scale.

Appendix C

Lagrange multipliers

Let $P(b)$ be the optimization problem

$$\begin{array}{ll} \text{minimize} & f(x) \\ \text{subject to} & h(x) = b \\ \text{over} & x \in X. \end{array}$$

Let $X(b) = \{x \in X : h(x) = b\}$. Say that x is *feasible* for $P(b)$ if $x \in X(b)$. Define the *Lagrangian*

$$L(x; y) = f(x) + y^T(b - h(x)).$$

Typically $X \subset \mathbb{R}^n$, $h : \mathbb{R}^n \to \mathbb{R}^m$, with $b, y \in \mathbb{R}^m$. The components of y are called *Lagrange multipliers*.

Theorem C.1 (Lagrange sufficiency theorem) *If $\bar{x}$ and $\bar{y}$ exist such that $\bar{x}$ is feasible for $P(b)$ and*

$$L(\bar{x}; \bar{y}) \leq L(x; \bar{y}), \qquad \forall x \in X,$$

then $\bar{x}$ is optimal for $P(b)$.

Proof For all $x \in X(b)$ and y we have

$$f(x) = f(x) + y^T(b - h(x)) = L(x; y).$$

Now $\bar{x} \in X(b) \subset X$, and thus

$$f(\bar{x}) = L(\bar{x}; \bar{y}) \leq L(x; \bar{y}) = f(x), \qquad \forall x \in X(b).$$

□

Section 2.7 gives an example of the use of a Lagrange multiplier, and of the application of the above theorem.

Let

$$\phi(b) = \inf_{x \in X(b)} f(x) \quad \text{and} \quad g(y) = \inf_{x \in X} L(x; y).$$

Then, for all y,

$$\phi(b) = \inf_{x \in X(b)} L(x; y) \geq \inf_{x \in X} L(x; y) = g(y). \tag{C.1}$$

Let $Y = \{y : g(y) > -\infty\}$. From inequality (C.1), we immediately have the following.

Theorem C.2 (Weak duality theorem)

$$\inf_{x \in X(b)} f(x) \geq \sup_{y \in Y} g(y).$$

The left-hand side of this inequality, that is $P(b)$, is called the *primal problem*, and the right-hand side is called the *dual problem*. The Lagrange multipliers y appearing in the dual problem are also known as *dual variables*. In general, the inequality in Theorem C.2 may be strict: in this case there is no $\bar{y}$ that allows the Lagrange sufficiency theorem to be applied.

Say that the *strong Lagrangian principle* holds for $P(b)$ if there exists $\bar{y}$ such that

$$\phi(b) = \inf_{x \in X} L(x; \bar{y});$$

equality then holds in Theorem C.2, from (C.1), and we say there is *strong duality*.

What are sufficient conditions for the strong Lagrangian principle to hold? We'll state, but not prove, some sufficient conditions that are satisfied for the various optimization problems we have seen in earlier chapters (for more, see Whittle (1971); Boyd and Vandenberghe (2004)).

Theorem C.3 (Supporting hyperplane theorem) *Suppose ϕ is convex and b lies in the interior of the set of points where ϕ is finite. Then there exists a (non-vertical) supporting hyperplane to ϕ at b.*

Theorem C.4 *The following are equivalent for the problem $P(b)$:*

- *there exists a (non-vertical) supporting hyperplane to ϕ at b;*
- *the strong Lagrangian principle holds.*

Corollary C.5 *If the strong Lagrangian principle holds, and if $\phi(\cdot)$ is differentiable at b, then $\bar{y} = \phi'(b)$.*

Remark C.6 The optimal value achieved in problem $P(b)$ is $\phi(b)$, and thus Corollary C.5 shows that the Lagrange multipliers $\bar{y}$ measure the sensitivity of the optimal value to changes in the constraint vector b. For this reason, Lagrange multipliers are sometimes called *shadow prices*.

Sometimes our constraints are of the form $h(x) \leq b$ rather than $h(x) = b$. In this case, we can write the problem in the form $P(b)$ as

$$\begin{array}{ll} \text{minimize} & f(x) \\ \text{subject to} & h(x) + z = b \\ \text{over} & x \in X, \ z \geq 0, \end{array} \tag{C.2}$$

where z is a vector of *slack variables*. Note that x, z now replaces x, and we look to minimize

$$L(x, z; y) = f(x) + y^T(b - h(x) - z)$$

over $x \in X$ and $z \geq 0$. For this minimization to give a finite value, it is necessary that $y \geq 0$, which is thus a constraint on the feasible region Y of the dual problem; and given this, any minimizing z satisfies $y^T z = 0$, a condition termed *complementary slackness*. Section 3.4 uses this formulation.

Theorem C.7 *For a problem $P(b)$ of the form* (C.2), *if X is a convex set, and f and h are convex, then ϕ is convex.*

Thus, if X is a convex set, f and h are convex functions, and b lies in the interior of the set of points where ϕ is finite, we conclude that the strong Lagrangian principle holds.

Appendix D

Foster–Lyapunov criteria

Our interest here is to be able to prove that certain Markov chains are positive recurrent without explicitly deriving the equilibrium distribution for them. We borrow an idea from the analysis of ordinary differential equations.

In Section 7.3, we studied the long-term behaviour of the trajectories of systems of ordinary differential equations. We were able to establish convergence without explicitly computing the trajectories themselves by exhibiting a Lyapunov function, i.e. a scalar function of the state of the system that was monotone in time.

A similar argument works for Markov chains. Consider an irreducible countable state space Markov chain $(X_n)_{n\geq 0}$. Analogous to the differential equation case, a Lyapunov function for a Markov chain is a function of the state X whose expectation is monotone in time: $\mathbb{E}[\mathcal{L}(X(n+1))\,|\,X(n)] \leq \mathcal{L}(X(n))$. However, as we noted in Chapter 5 (Remark 5.2), a simple drift condition is not enough to show positive recurrence; we need some further constraints bounding the upward jumps of $\mathcal{L}$.

Proposition D.1 (Foster–Lyapunov stability criterion for Markov chains) *Let $(X(n))_{n\geq 0}$ be an irreducible Markov chain with state space $\mathcal{S}$ and transition matrix $P = (p(i,j))$. Suppose $\mathcal{L} : \mathcal{S} \to \mathbb{R}_+$ is a function such that, for some constants $\epsilon > 0$ and b, some finite exception set $K \subset \mathcal{S}$, and all $i \in \mathcal{S}$,*

$$\mathbb{E}[\mathcal{L}(X(n+1)) - \mathcal{L}(X(n))\,|\,X(n) = i] \leq \begin{cases} -\epsilon, & i \notin K, \\ b - \epsilon, & i \in K. \end{cases}$$

Then the expected return time to K is finite, and X is a positive recurrent Markov chain.

Proof Let $\tau = \min\{n \geq 1 : X(n) \in K\}$ be the first time after $t = 0$ of entering the exception set. We show that $\mathbb{E}[\tau\,|\,X(0) = i]$ is finite for all $i \in \mathcal{S}$.

Consider the inequality

$$\mathbb{E}[\mathcal{L}(X(t+1)) \mid X(t)] + \epsilon \le \mathcal{L}(X(t)) + bI\{X(t) \in K\},$$

which holds for all times t. Add this up over all $t : 0 \le t \le \tau - 1$ and take expectations to obtain

$$\sum_{t=1}^{\tau} \mathbb{E}[\mathcal{L}(X(t))] + \epsilon\mathbb{E}[\tau] \le \sum_{t=0}^{\tau-1} \mathbb{E}[\mathcal{L}(X(t))] + b. \tag{D.1}$$

We would like to cancel like terms in the sums to get a bound on $\mathbb{E}[\tau]$, but the sums may be infinite. We get around this difficulty as follows. Let τ^n be the first time when $\mathcal{L}(X(t)) \ge n$, and let

$$\tau_n = \min(\tau, n, \tau^n).$$

Then we have

$$\sum_{t=1}^{\tau_n} \mathbb{E}[\mathcal{L}(X(t))] + \epsilon\mathbb{E}[\tau_n] \le \sum_{t=0}^{\tau_n-1} \mathbb{E}[\mathcal{L}(X(t))] + b, \tag{D.2}$$

and here everything is finite, so we may indeed cancel like terms:

$$\epsilon\mathbb{E}[\tau_n] \le \mathcal{L}(X(0)) - \mathcal{L}(X(\tau_n)) + b \le \mathcal{L}(X(0)) + b. \tag{D.3}$$

Increasing $n \to \infty$ will increase $\tau^n \to \infty$, hence $\tau_n \to \tau$. Monotone convergence now implies

$$\mathbb{E}[\tau] \le \frac{\mathcal{L}(X(0)) + b}{\epsilon}.$$

We have thus shown that the mean return time to the exception set K is finite. This implies (and is equivalent to) positive recurrence (Asmussen (2003), Lemma I.3.10). □

Remark D.2 This technique for proving stability of Markov chains was introduced in Foster (1953). Many variants of the Foster–Lyapunov criteria exist; this version has been adapted from Hajek (2006).

Often we want to work with (continuous time) Markov processes, rather than (discrete time) Markov chains. The Foster–Lyapunov stability criterion is very similar, once we have defined the drift appropriately. Looking over a small time interval of length δ, we have

$$\frac{\mathbb{E}[\mathcal{L}(X(t+\delta)) - \mathcal{L}(X(t)) \,|\, X(t) = i]}{\delta}$$
$$= \sum_j \underbrace{\frac{\mathbb{P}(X(t+\delta) = j \,|\, X(t) = i))}{\delta}}_{\to q(i,j)} (\mathcal{L}(j) - \mathcal{L}(i))$$

as $\delta \to 0$, where $(q(i, j))$ is the matrix of transition rates for the Markov process. Thus, we expect the drift condition for Markov processes to use the quantity $\sum_j q(i, j)(\mathcal{L}(j) - \mathcal{L}(i))$.

Proposition D.3 (Foster–Lyapunov stability criteria for Markov processes) *Let $(X(t))_{t\geq 0}$ be a (time-homogeneous, irreducible, non-explosive, conservative) continuous time Markov process with countable state space $\mathcal{S}$ and matrix of transition rates $(q(i, j))$. Suppose $\mathcal{L} : \mathcal{S} \to \mathbb{R}_+$ is a function such that, for some constants $\epsilon > 0$ and b, some finite exception set $K \subset \mathcal{S}$, and all $i \in \mathcal{S}$,*

$$\sum_j q(i, j)(\mathcal{L}(j) - \mathcal{L}(i)) \leq \begin{cases} -\epsilon, & i \notin K, \\ b - \epsilon, & i \in K. \end{cases} \tag{D.4}$$

Then the expected return time to K is finite, and X is positive recurrent.

Proof Let X^J be the jump chain of X. This is the Markov chain obtained by looking at X just after it has jumped to a new state. Recall, from Chapter 1, that the transition probabilities of the jump chain are $p(i, j) = q(i, j)/q(i)$, where $q(i) = \sum_j q(i, j)$. Let $N = \min\{n \geq 1 : X^J(n) \in K\}$, and let τ be the time of the Nth jump. Equivalently, τ is the first time after leaving state $X(0)$ that the process $X(t)$ is in the set K.

Now, we can rewrite the condition (D.4) as

$$\sum_j p(i, j)(\mathcal{L}(j) - \mathcal{L}(i)) \leq \begin{cases} -\tilde{\epsilon}(i) = -\epsilon q(i)^{-1}, & i \notin K, \\ \tilde{b}(i) - \tilde{\epsilon}(i) = (b - \epsilon) q(i)^{-1}, & i \in K. \end{cases}$$

The argument for discrete time chains gives

$$\mathbb{E}\left[\sum_{n=0}^{N-1} \tilde{\epsilon}(X^J(n))\right] \leq \mathcal{L}(X(0)) + \tilde{b}(X(0)).$$

Recall that $q(i)^{-1}$ is the expected time it takes the Markov process X to jump from state i, so the above inequality can be rewritten as

$$\epsilon\mathbb{E}[\tau] \leq \mathcal{L}(X(0)) + \tilde{b}(X(0)),$$

and hence $\mathbb{E}[\tau] < \infty$. A continuous time version of (Asmussen (2003) lemma I.3.10) then implies that X is positive recurrent. □

Foster–Lyapunov criteria provide a powerful technique for showing positive recurrence of Markov chains, because constructing a Lyapunov function is often easier than constructing an explicit, summable equilibrium distribution. However, verifying the conditions of the criteria can be a lengthy process. An example computation from Section 7.7.8 appears in Example D.4 and Exercise D.1.

The question of determining when queueing systems are stable is an active area of research. Lyapunov functions offer one technique, but they come with limitations, both because of the Markovian assumptions required, and because writing down an explicit Lyapunov function can be tricky – see Exercise D.3. One commonly adapted approach involves constructing *fluid limits*, which describe the limiting trajectories of the system under a certain rescaling. Analyzing properties of these trajectories is often simpler than constructing an explicit Lyapunov function, and can also be used to show positive recurrence. To read about this technique, see Bramson (2006).

Example D.4 (MaxWeight-α) Recall the model of a switched network running the MaxWeight-α algorithm. There is a set of queues indexed by $\mathcal{R}$. Packets arrive into queues as independent Poisson processes with rates λ_r. Due to constraints on simultaneous activation of multiple queues, at each discrete time slot there is a finite set $\mathcal{S}$ of possible schedules, only one of which can be chosen. A schedule σ is a vector of non-negative integers $(\sigma_r)_{r\in\mathcal{R}}$, which describes the number of packets to be served from each queue. (If $q_r < \sigma_r$, i.e. there aren't enough packets in queue r, the policy will serve all the packets in queue r.) The MaxWeight-α policy will, at every time slot, pick a schedule σ that solves

$$\text{maximize} \sum_{r\in\mathcal{R}} \sigma_r q_r(t)^\alpha$$
$$\text{subject to } \sigma \in \mathcal{S}.$$

Note that this is a Markovian system, since the scheduling decision depends only on the current state of the system. We would like to know for what arrival rate vectors λ the Markov chain is positive recurrent. Theorem 7.13 asserts that this is true for all λ in the interior of the admissible region Λ, given by

$$\Lambda = \left\{ \lambda \geq 0 : \exists c_\sigma \geq 0, \sigma \in \mathcal{S}, \sum_\sigma c_\sigma = 1 \text{ such that } \lambda \leq \sum_\sigma c_\sigma \sigma \right\}.$$

In our sketch of the proof of this result, we considered the Lyapunov function

$$\mathcal{L}(q) = \sum \frac{q_r^{\alpha+1}}{\alpha + 1}.$$

In our analysis in Chapter 7, we approximated its true drift by a quantity that was easier to work with; see (7.14). Let us now show rigorously that this function $\mathcal{L}$ satisfies the conditions of Proposition D.1.

We will show that the drift of $\mathcal{L}(q(t))$ can be bounded as follows:

$$\begin{aligned}\mathbb{E}[\mathcal{L}(q(t+1)) - \mathcal{L}(q(t)) \,|\, q(t)] \leq & \underbrace{\sum_r q_r(t)^\alpha \mathbb{E}[q_r(t+1) - q_r(t) \,|\, q_r(t)]}_{\text{main term}} \\ & + \underbrace{\delta \sum_r q_r(t)^\alpha + \sum_r b_r}_{\text{error}},\end{aligned} \tag{D.5}$$

where we can take constants $\delta_r > 0$ to be arbitrarily small, and then pick constants b_r to be sufficiently large. The main term can be bounded above by $-\epsilon\|q(t)\|^r$ as in Exercise 7.18, for some $\epsilon > 0$. Clearly, if δ_r are small enough, the main term will dominate for large enough queues. Thus, we can pick a constant $K > 0$ for which the conditions of Proposition D.1 are satisfied.

It remains to show the bound (D.5). You will do this in Exercise D.1.

Exercises

Exercise D.1 Write $\mathbb{E}[q_r(t+1)^{\alpha+1} - q_r(t)^{\alpha+1} \,|\, q(t)]$ as

$$q_r(t)^{\alpha+1} \mathbb{E}\left[\left(\left(1 + \frac{q_r(t+1) - q_r(t)}{q_r(t)}\right)^{\alpha+1} - 1\right) \,|\, q(t)\right].$$

The (random) increment $|q_r(t+1) - q_r(t)|$ is bounded above, using the maximum of a Poisson random variable of rate λ_r (the arrivals) and the finite set of values σ_r (the services). Call this maximum Y_r, and note that Y_r is independent of $q_r(t)$.

(1) Show that we can bound the drift above by

$$q_r(t)^{\alpha+1}\mathbb{E}\left[\left(\left(1+\frac{Y_r}{q_r(t)}\right)^{\alpha+1}-1\right) \mid q_r(t)\right].$$

(2) Let K_r be a large constant, let $\delta > 0$ be small, and suppose $q_r(t) > K_r$. By splitting the expectation (over Y) into the cases of $Y > \delta K_r$ and $Y \leq \delta K_r$, show that

$$\mathbb{E}\left[\left(\left(1+\frac{Y_r}{q_r(t)}\right)^{\alpha+1}-1\right) \mid q_r(t)\right]$$
$$\leq \left|(1+\delta)^{\alpha+1}-1\right| + \mathbb{E}\left[\left|\left(1+\frac{Y_r}{K_r}\right)^{\alpha+1}\right| + 1 \mid Y_r > \delta K_r\right]\mathbb{P}(Y_r > \delta K_r).$$

(3) Show that all moments of Y_r are finite. Conclude that the second term tends to zero as $K_r \to \infty$.

(4) If $q_r(t) \leq K_r$, show that the right-hand side is bounded by a constant b_r (which may depend on K_r).

(5) Show that

$$\mathbb{E}\left[\frac{1}{\alpha+1}\left(q_r(t+1)^{\alpha+1} - q_r(t)^{\alpha+1}\right) \mid q(t)\right]$$
$$\leq q_r(t)^{\alpha}\mathbb{E}[q_r(t+1) - q_r(t) \mid q_r(t)] + 2\delta\sum_r q_r(t)^{\alpha} + \sum_r b_r,$$

provided δ is chosen small enough, then K_r large enough, then b_r large enough.

Exercise D.2 Perform a similar analysis to that in Exercise D.1 for the α-fair allocation of Section 8.3. The Lyapunov function is defined in the proof of Theorem 8.2.

Exercise D.3 In this exercise, we revisit the backlog estimation scheme considered in Section 5.2, and show that it is positive recurrent for $\nu < e^{-1}$ and a choice of parameters $(a, b, c) = (2 - e, 0, 1)$. Finding the Lyapunov function is quite tricky here; the suggestion below is adapted from (Hajek (2006), Proof of proposition 4.2.1).

Recall that $\kappa_\nu < 1$ was defined as the limiting value of $\kappa(t) = n(t)/s(t)$ for a trajectory of the system of differential equations approximating the Markov chain $(N(t), S(t))$ (see equations (5.3)). To construct the Lyapunov function, we observe that for the differential equation either $n(t) \approx \kappa_\nu s(t)$, and then $n(t)$ is decreasing, or else the deviation $|n(t) - \kappa_v s(t)|$ tends to decrease. Thus, the Lyapunov function below includes contributions from both of those quantities.

Define

$$\mathcal{L}(N, S) = N + \phi(|\kappa_v S - N|), \quad \phi(u) = \begin{cases} u^2/M, & 0 \le u \le M^2, \\ M(2u - M^2), & u > M^2, \end{cases} \tag{D.6}$$

for some sufficiently large constant M.

Show that $\mathcal{L}(n(t), s(t))$ is monotonic when $(n(t), s(t))$ solve (5.3). Then use the techniques in this appendix to argue that $\mathcal{L}$ satisfies the conditions of the Foster–Lyapunov criteria for the Markov chain (N_t, S_t).

References

Aldous, D. 1987. Ultimate instability of exponential back-off protocol for acknowledgement-based transmission control of random access communication channels. *IEEE Transactions on Information Theory*, **33**, 219–223.

Asmussen, S. 2003. *Applied Probability and Queues*. 2nd edn. New York, NY: Springer.

Baccelli, F. and Brémaud, P. 2003. *Elements of Queueing Theory*. Berlin: Springer.

BenFredj, S., Bonald, T., Proutière, A., Régnié, G. and Roberts, J. W. 2001. Statistical bandwidth sharing: a study of congestion at flow level. *Computer Communication Review*, **31**, 111–122.

Berry, R. A. and Johari, R. 2013. *Economic Modeling: A Primer with Engineering Applications*. Foundations and Trends in Networking. Delft: now publishers.

Bonald, T. and Massoulié, L. 2001. Impact of fairness on Internet performance. *Performance Evaluation Review*, **29**, 82–91.

Boyd, S. and Vandenberghe, L. 2004. *Convex Optimization*. Cambridge: Cambridge University Press.

Bramson, M. 2006. *Stability and Heavy Traffic Limits for Queueing Networks: St. Flour Lecture Notes*. Berlin: Springer.

Chang, C.-S. 2000. *Performance Guarantees in Communication Networks*. London: Springer.

Chen, M., Liew, S., Shao, Z. and Kai, C. 2013. Markov approximation for combinatorial network optimization. *IEEE Transactions on Information Theory*. doi: 10.1109/TIT.2013.2268923.

Chiang, M., Low, S. H., Calderbank, A. R. and Doyle, J. C. 2007. Layering as optimization decomposition: a mathematical theory of network architectures. *Proceedings of the IEEE*, **95**, 255–312.

Courcoubetis, C. and Weber, R. 2003. *Pricing Communication Networks: Economics, Technology and Modelling*. Chichester: Wiley.

Crametz, J.-P. and Hunt, P. J. 1991. A limit result respecting graph structure for a fully connected loss network with alternative routing. *Annals of Applied Probability*, **1**, 436–444.

Crowcroft, J. and Oechslin, P. 1998. Differentiated end-to-end Internet services using a weighted proportionally fair sharing TCP. *Computer Communications Review*, **28**, 53–69.

Doyle, P. G. and Snell, J. L. 2000. *Random Walks and Electric Networks*. Carus Mathematical Monographs. Washington, D.C.: The Mathematical Association of America.

Erlang, A. K. 1925. A proof of Maxwell's law, the principal proposition in the kinetic theory of gases. In Brockmeyer, E., Halstrom, H. L. and Jensen, A. (eds.), *The Life and Works of A. K. Erlang*. Copenhagen: Academy of Technical Sciences, 1948, pp. 222–226.

Foster, F. G. 1953. On the stochastic matrices associated with certain queueing processes. *Annals of Mathematical Statistics*, **24**, 355–360.

Gale, D. 1960. *The Theory of Linear Economic Models*. Chicago, IL: The University of Chicago Press.

Gallager, R. G. 1977. A minimum delay routing algorithm using distributed computation. *IEEE Transactions on Communications*, **25**, 73–85.

Ganesh, A., O'Connell, N. and Wischik, D. 2004. *Big Queues*. Berlin: Springer.

Gibbens, R. J., Kelly, F. P. and Key, P. B. 1995. Dynamic alternative routing. In Steenstrup, Martha (ed.), *Routing in Communications Networks*. Englewood Cliffs, NJ: Prentice Hall, pp. 13–47.

Goldberg, L., Jerrum, M., Kannan, S. and Paterson, M. 2004. A bound on the capacity of backoff and acknowledgement-based protocols. *SIAM Journal on Computing*, **33**, 313–331.

Hajek, B. 2006. *Notes for ECE 467: Communication Network Analysis*. `http://www.ifp.illinois.edu/~hajek/Papers/networkanalysis.html`.

Jacobson, V. 1988. Congestion avoidance and control. *Computer Communication Review*, **18**, 314–329.

Jiang, L. and Walrand, J. 2010. A distributed CSMA algorithm for throughput and utility maximization in wireless networks. *IEEE/ACM Transactions on Networking*, **18**, 960–972.

Jiang, L. and Walrand, J. 2012. Stability and delay of distributed scheduling algorithms for networks of conflicting queues. *Queueing Systems*, **72**, 161–187.

Johari, R. and Tsitsiklis, J. N. 2004. Efficiency loss in a network resource allocation game. *Mathematics of Operations Research*, **29**, 407–435.

Kang, W. N., Kelly, F. P., Lee, N. H. and Williams, R. J. 2009. State space collapse and diffusion approximation for a network operating under a fair bandwidth-sharing policy. *Annals of Applied Probability*, **19**, 1719–1780.

Kelly, F. P. 1991. Loss networks. *Annals of Applied Probability*, **1**, 319–378.

Kelly, F. P. 1996. Notes on effective bandwidths. In Kelly, F. P., Zachary, S. and Ziedins, I.B. (eds.), *Stochastic Networks: Theory and Applications*. Oxford: Oxford University Press, pp. 141–168.

Kelly, F. P. 2003a. Fairness and stability of end-to-end congestion control. *European Journal of Control*, **9**, 159–176.

Kelly, F. P. 2011. *Reversibility and Stochastic Networks*. Cambridge: Cambridge University Press.

Kelly, F. P. and MacPhee, I. M. 1987. The number of packets transmitted by collision detect random access schemes. *Annals of Probability*, **15**, 1557–1668.

Kelly, F. P. and Raina, G. 2011. Explicit congestion control: charging, fairness and admission management. In Ramamurthy, B., Rouskas, G. and Sivalingam, K. (eds.), *Next-Generation Internet Architectures and Protocols*. Cambridge: Cambridge University Press, pp. 257–274.

Kelly, T. 2003b. Scalable TCP: improving performance in highspeed wide area networks. *Computer Communication Review*, **33**, 83–91.

Kendall, D. G. 1953. Stochastic processes occurring in the theory of queues and their analysis by the method of the imbedded Markov chain. *Annals of Mathematical Statistics*, **24**, 338–354.

Kendall, D. G. 1975. Some problems in mathematical genealogy. In Gani, J. (ed.), *Perspectives in Probability and Statistics: Papers in Honour of M.S. Bartlett*. London: Applied Probability Trust/Academic Press, pp. 325–345.

Key, P. B. 1988. Implied cost methodology and software tools for a fully connected network with DAR and trunk reservation. *British Telecom Technology Journal*, **6**, 52–65.

Kind, J., Niessen, T. and Mathar, R. 1998. Theory of maximum packing and related channel assignment strategies for cellular radio networks. *Mathematical Methods of Operations Research*, **48**, 1–16.

Kingman, J. F. C. 1993. *Poisson Processes*. Oxford: Oxford University Press.

Kleinrock, L. 1964. *Communication Nets: Stochastic Message Flow and Delay*. New York, NY: McGraw Hill.

Kleinrock, L. 1976. *Queueing Systems, vol II: Computer Applications*. New York, NY: Wiley.

Lu, S. H. and Kumar, P. R. 1991. Distributed scheduling based on due dates and buffer priorities. *IEEE Transactions on Automatic Control*, **36**, 1406–1416.

Marbach, P., Eryilmaz, A. and Ozdaglar, A. 2011. Asynchronous CSMA policies in multihop wireless networks with primary interference constraints. *IEEE Transactions on Information Theory*, **57**, 3644–3676.

Mazumdar, R. 2010. *Performance Modeling, Loss Networks, and Statistical Multiplexing*. San Rafael, CA: Morgan and Claypool.

Meyn, S. P. and Tweedie, R. L. 1993. *Markov Chains and Stochastic Stability*. London: Springer.

Moallemi, C. and Shah, D. 2010. On the flow-level dynamics of a packet-switched network. *Performance Evaluation Review*, **38**, 83–94.

Modiano, E., Shah, D. and Zussman, G. 2006. Maximizing throughput in wireless networks via gossiping. *Performance Evaluation Review*, **34**, 27–38.

Nash, J. F. 1950. The bargaining problem. *Econometrica*, **18**, 155–162.

Norris, J. R. 1998. *Markov Chains*. Cambridge: Cambridge University Press.

Ott, T. J. 2006. Rate of convergence for the 'square root formula' in the Internet transmission control protocol. *Advances in Applied Probability*, **38**, 1132–1154.

Pallant, D. L. and Taylor, P. G. 1995. Modeling handovers in cellular mobile networks with dynamic channel allocation. *Operations Research*, **43**, 33–42.

Pitman, J. 2006. *Combinatorial Stochastic Processes*. Berlin: Springer.

Rawls, J. 1971. *A Theory of Justice*. Cambridge, MA: Harvard University Press.

Ross, K. W. 1995. *Multiservice Loss Models for Broadband Communication Networks*. London: Springer.

Shah, D. and Shin, J. 2012. Randomized scheduling algorithm for queueing networks. *Annals of Applied Probability*, **22**, 128–171.

Shah, D. and Wischik, D. 2012. Switched networks with maximum weight policies: fluid approximation and multiplicative state space collapse. *Annals of Applied Probability*, **22**, 70–127.

Shah, D., Walton, N. S. and Zhong, Y. 2012. Optimal queue-size scaling in switched networks. *Performance Evaluation Review*, **40**, 17–28.

Shakkottai, S. and Srikant, R. 2007. *Network Optimization and Control.* Foundations and Trends in Networking. Hanover, MA: now publishers.

Songhurst, D. J. 1999. *Charging Communication Networks: From Theory to Practice.* Amsterdam: Elsevier.

Srikant, R. 2004. *The Mathematics of Internet Congestion Control.* Boston, MA: Birkhäuser.

Tassiulas, L. and Ephremides, A. 1992. Stability properties of constrained queueing systems and scheduling policies for maximum throughput in multihop radio networks. *IEEE Transactions on Automatic Control*, **37**, 1936–1948.

Tian, Y.-P. 2012. *Frequency-Domain Analysis and Design of Distributed Control Systems.* Singapore: Wiley.

Vinnicombe, G. 2002. On the stability of networks operating TCP-like congestion control. *Proc. 15th Int. Fed. Automatic Control World Congress, Barcelona, Spain*, 217–222.

Walrand, J. 1988. *An Introduction to Queueing Networks.* Englewood Cliffs, NJ: Prentice Hall.

Walton, N. S. 2009. Proportional fairness and its relationship with multi-class queueing networks. *Annals of Applied Probability*, **19**, 2301–2333.

Whittle, P. 1971. *Optimization Under Constraints.* New York, NY: Wiley.

Whittle, P. 1986. *Systems in Stochastic Equilibrium.* New York, NY: Wiley.

Whittle, P. 2007. *Networks: Optimisation and Evolution.* Cambridge: Cambridge University Press.

Wischik, D. 1999. The output of a switch, or, effective bandwidths for networks. *Queueing Systems*, **32**, 383–396.

Wischik, D., Raiciu, C., Greenhalgh, A. and Handley, M. 2011. Design, implementation and evaluation of congestion control for multipath TCP. *Proc. 8th USENIX Conference on Networked Systems Design and Implementation, Boston, MA*, 99–112.

Zachary, S. and Ziedins, I. 2011. Loss networks. In Boucherie, R.J. and van Dijk, N.M. (eds.), *Queueing Networks: A Fundamental Approach.* New York, NY: Springer, pp. 701–728.

Index